LISA SAMSON is an [...] [...]e specialist with many yea[...] [...]English and Italian. She was also a Senior Lecturer in Writing at Leeds Beckett University. Lisa's first novel, *Talk To Me*, came second in the Virginia Prize for Fiction 2011. She has been published in short form both in print and online.

Praise for *Epitaph for the Ash*:

'Fascinating … Her pilgrimage to discover the present state of the ash in the UK, and the work that is being done to accommodate or counter ash dieback, is both a labour of love and an extraordinary achievement, especially given the heart-rending physical limitations Samson eventually endures as a result of life-saving surgery'   *Observer*

'Everywhere Lisa's powerful affinity with the natural world is palpable … Samson is right to urge appreciation of what we have; her book will also help raise awareness of the need to protect our invaluable natural heritage for future generations'   *Literary Review*

'One of the absolute hardwood strengths of *Epitaph for the Ash* is its author's appreciation of *Fraxinus excelsior* in British history'   *Country Life*

'Lisa Samson's … quest to travel the length and breadth of the land takes us on a leafy green jewel of a journey into a kingdom that will change the way you look at the ash tribe forever' *BBC Wildlife*

Epitaph for the Ash

In Search of Recovery and Renewal

Lisa Samson

4th ESTATE • London

4th Estate
An imprint of HarperCollins*Publishers*
1 London Bridge Street
London SE1 9GF

www.4thEstate.co.uk

First published in Great Britain in 2018 by 4th Estate
This 4th Estate paperback edition 2019

1

A catalogue record for this book is
available from the British Library

ISBN 978-0-00-754463-9

Printed and bound in Great Britain by
CPI Group (UK) Ltd, Croydon, CR0 4YY

MIX
Paper from
responsible sources
FSC® C007454

This book is produced from independently certified FSC paper
to ensure responsible forest management

For more information visit: www.harpercollins.co.uk/green

For Mike, without whom this book
wouldn't have been possible

Contents

Introduction   1

The Blight of Ashwellthorpe   5

The Science behind Ash Dieback   25

Secrets   35

Bread   53

Ancestors   69

Ash Fires   95

Life Cycle   115

Spring   123

Isolation   145

Shadows   167

Shelter   177

Resistance   199

Notes   207

Acknowledgements   215

Introduction

You may walk past the ash – its slenderness and height blend easily into any wood or hedgerow – but in spring you'll stop to admire the bluebells shimmering in the light that filters through its foliage. The continued existence of the ash tree is under threat from Ash Dieback, a disease that has spread from the Continent and is threatening ash trees in Britain. During the research and writing for this book, I was diagnosed with a benign brain tumour, which changed my life irrevocably, never to be the same again. *Epitaph for the Ash* is a celebration of the ash and an account of my personal journey as I recorded its decline over the last few years, taking the reader from the lowlands of Norfolk to the uplands of Yorkshire, and from Devon to the northernmost reaches of the British Isles. The book explores the cultural significance of the ash tree, tracing its mythology in Norse culture and through some of the literature on and history of woodlands in Britain.

The ancient woodlands of today are a fraction of the size they were in Anglo-Saxon Britain, as agriculture and industry have gradually encroached on the forests. During the post-war years thousands of acres of woodland were given over to agricultural production or buildings to accommodate a burgeoning population. As the elms were ravaged by Dutch Elm Disease in the latter half of the twentieth century, the importance of the ash as a habitat for rare flora and fauna increased. Trees, like any living organism, have always suffered from blight and disease, but the loss of most of our elms, and now the danger to the ash poses a serious environmental threat.

In May 1978, I clearly recall my mother bristling with pride as she showed me a copy of the *Sunday Times Magazine*, which featured my uncle's new book, *Epitaph for the Elm*. Gerald Wilkinson was her older brother, the eccentric tree expert. We saw little of him, but occasionally he would stop by in his old Volkswagen camper-van on his way to or from one of his extended research trips in the woodlands of Britain. To a town child like me, with a yearning for the open countryside, this seemed the most romantic way to spend your life, and I'm sure I begged him to take me with him.

I've been communing with trees since I was young and sometimes fancy they are aware of me. The brightest days of my childhood were when my family and I got onto a bus into the countryside, then walked through the woods and fields, picnicking under a tree. I'd roam away from the others and listen to the trees whispering. I never wandered too far because I was afraid of my own shadow and would imagine danger lurking behind hedgerows and in the depths of the dark wood. I've always loved trees for their swaying limbs and shady canopies, so easy to draw, but it is only as an adult that I have learned to regard them as friends. They're good company and they do talk if you stop to listen.

I was too young to understand the full impact of Dutch Elm Disease, but as I grew up and my own interest in trees deepened, I often referred to Gerald's books. His death in a car crash in 1988 fuelled my intention to learn more about his obsession with them. Now, nearly three decades later, it is predicted that *Chalara fraxinea*, the fungus that causes *Hymenoscyphus fraxineus* or *Hymenoscyphus pseudoalbidus* – Ash Dieback – will ravage the ash trees, again changing Britain's landscape. It was found in the wild in Britain in 2012, in Ashwellthorpe, Norfolk, and since then it has spread relatively quickly, with new cases reported often in the press. My home county of Yorkshire will seem barren without the ancient ashes protruding from limestone scars and chalky cliff faces, or spreading their fine canopies over the hedgerows.

The ash's status as a 'magic' tree with healing properties gives it a fascinating history. Gerald suggests that in Neolithic times 'the Ash may have been sacred'. Druids regard it as such, and in Norse mythology the ash was the Tree of Life, the most important living thing besides humans. It is one of our strongest trees, used for framework in vehicles and tool handles, but craftspeople and manufacturers are already using other materials.

Across the length and breadth of Britain place names are associated with the ash – Ash, Ashburton, Ashby-de-la-Zouch, Ashbourne, Ash-cum-Ridley, Ashover, Ashwater, Ashtead, Ashkirk, Ashcot, Ashwell – anthropological mnemonics linking people to their places. Some will have been named after self-sown ash seeds that blew on the wind, others for the ash their inhabitants cultivated, but all bear testimony to the part the ash has played in our civilization.

The Blight of Ashwellthorpe

My first view of Ashwellthorpe is of a bright yellow rape field where poppies blush under a louring sky, June, 2013. The wood comes into view, 'a darker green than usual', like the Enchanted Wood of Enid Blyton's books. It is dense with many species of tree. In rich full leaf, they nod and sway in the dull light, waiting for the rain. They beckon to me over the hedgerows and houses, but the road running straight through the village leads me past. I nearly hit the kerb as I try to glimpse the ashes I've come all this way to see. An elderly cyclist ahead of me pauses to let me pass, perhaps to save himself or to stare at me, a stranger in this sleepy village. I pass the hall and the church, the rows of plain houses and cottages that line the road: a barrier obstructing my view of the wood.

The village sweeps by and I pull up in the lane beside three ashes that all have the characteristic antlers of leafless upper branches, like oak trees widening their canopy as they mature. Crown dieback is a symptom of *Chalara fraxinea*'s presence and these trees remind me of photos I've seen of diseased trees in Poland. It could be part of a natural process ... or signs of Ash Dieback. Naturally, I go for the latter theory since I know it has spread widely across the Norfolk Broads.

I climb out of my car and look back at the village and the wood. It is low-lying, cradled in the embrace of the fields and village that have depended on it for centuries. There is a gap between a larger and smaller patch of woodland, mown through as straight as a Roman road. The smaller wood is Upper Wood and the larger is Lower Wood. They are believed to have been separated during the Napoleonic Wars in the early 1800s, when a perfectly straight line of vision was required between London and the Norfolk coast for semaphore messages to be relayed as part of an early-warning system. Apparently it took only half an hour to relay the message of an approaching ship near Great Yarmouth all the way to London. Upper Wood is now privately owned and is a jealously guarded pheasant-shoot, but Lower Wood is managed by the Norfolk Wildlife Trust and is a site of designated specific interest. It is also the first place in Britain where symptoms of Ash Dieback were discovered in the wild.

The woods are beckoning to me. I fancy I can hear them whispering their secrets, like the trees in *The Magic Faraway Tree*. Blyton's enchanted tree is huge and guarded by fairy folk, its branches reaching into different magical worlds. Its roots stretch far into the earth and it represents a complete entity, both admired and feared by the woodland folk who live nearby. Although the Faraway Tree is of an unspecified type, Blyton was perhaps aware of the ash's magical properties and its powers of protection. It bears some resemblance to Yggdrasil, a huge tree that was at the centre of the Norse universe. Its roots were so abundant and so long that they reached into the underworld, and its trunk was so tall that its branches stretched to Heaven. Yggdrasil was the giver of life, at the beginning of all creation.

I drag myself back into my car and drive on to Norwich, knowing I will return in the late afternoon. Being among green things, in nature, is almost as essential to me as breathing: I cannot go too long without it. I need to feel the air, sun and rain on my face, to hear the wind blowing through the trees and grass. With a pang, I watch the woods receding from view and just refrain from stopping the car to go back.

*

It's pouring with rain when I turn into the tiny Rosemary Meadow car park in the late afternoon. Steve Collin, head forester with the Norfolk Wildlife Trust, has to flag me down. No bigger than a garden, it backs onto the Wildlife Trust Meadow that skirts the south side of the wood. It is bright with Greater Spearwort, like large buttercups, and tightly clustered Red Clover, but today the grasses hang their heads, beaten down by the rain, which has been gathering force all afternoon. Steve seems not to notice it as he squelches towards the wood, with me slithering behind. He is at home here and strides with an air of propriety, accustomed to leading people round the wood that has received so much media attention in recent months.

The dripping trees enfold us in their shadows, sheltering us from the worst of the downpour. Fat raindrops splay on my glasses and blur my vision as I look up to the top of the canopy to see the tall ashes that have shot above the roof of the wood in their eagerness to reach the light. At my level, five feet four from the ground, it would be easy to walk past the trunks of the young ash without noticing what they are, since their bark is darkened with damp. They sway in the gloom of the wet wood, their leaves rustling, sweeping away the rain.

Twigs snap underfoot as branches of alder and ash brush our legs and reach out to touch our shoulders. A short way into the woods, we stop by what is probably the most photographed tree of 2012: the ash sapling on which Steve first noticed signs of disease in October that year. It has a diamond-shaped lesion close to a shoot some way up its thin trunk, which is seeping resin, like congealed blood from an open wound. Unlike human flesh, though, the tree bark cannot heal. Steve breaks off a dead shoot to show me how brittle it is. The effects of *Chalara* are easier to spot on a thin young tree because the diseased girdle that forms around the trunk is visible and everything above it dies. Leaves have blackened and twigs have wilted, ready to fall to the ground, no use to anyone. Most of the dead leaves and wood are removed once they've fallen, as instructed by the Forestry Commission.

On the day they first spotted the disease, Steve Collin and Dr Anne Edwards, the volunteer warden at Ashwellthorpe Woods, were looking at the coppicing that had been done. After twenty years, they had finally brought the cycle of coppicing into a regular rotation and the woods were responding to their care, with an annual increase in the numbers of bluebells and wild orchids that would not thrive as well without the light exposed by coppicing. Steve and Anne noticed that some of the trees appeared to be dead and reported it to the Forestry Commission, suspecting an invasion of *Chalara fraxinea*.

Meanwhile Anne, a scientist, took samples back to the John Innes Centre where she works and alerted her colleagues to the potential danger lurking in their own backyard. Within weeks *Chalara* was confirmed but they knew they wouldn't be able to gauge the real level of damage until the autumn leaves had fallen and the new buds had unfurled on the trees in spring. We discuss whether the fungus was blown in spores across the sea from Denmark, and Steve notes that the wind has been blowing easterly for a few years. In addition, the heavy rains in recent summers made perfect breeding ground for the fungus, although the ashes themselves can thrive in wet conditions.

An area of saplings holds yellowed shoots that have been drained of life. A bright yellow one is still sprouting from the base, though most of the tree above the infected girdle is dead. It hides behind a group of healthy green shoots. The saplings' bark is slick with rain that trickles over a purplish gash, the mark of disease. Most of the signs of Ash Dieback in Ashwellthorpe are in the coppice growth, but the big ashes at the back of the coppice have Ash Dieback at the top and have probably had it for two or three years at least. The rooks perched on the uppermost branches are oblivious, croaking to one another in the dark heart of the wood.

Ash is the predominant species here but it is a mixed coppice of hazel, ash, hawthorn, alder and sallow. Ash, closely followed by alder, was possibly the commonest tree to make up the forests of the Neolithic wildwood that covered the Broadlands when the sheep of Neolithic peoples began to nibble at the tasty bark and seeds, turning the woods to scrub. Much of the woodland would have been felled to make way for farming and building. By the time the Anglo-Saxons, and later the Vikings, arrived, the woods were shrinking, with fields taking over. The Domesday Book of 1081 makes reference to a large tract of woodland here, and the present day Ashwellthorpe Woods are the last vestige of its former dimensions. The woods have retained their current size since about the 1830s. Seen from above on aerial photos, they form two darker blocks of green among the patches of yellow cereals, pale green wheat and brown ploughed land.

Behind Ashwellthorpe Hall, to the east of the wood, the remains of an Anglo-Saxon burial site were found, so it is safe to assume that there has been a settlement here since the early Anglo-Saxons arrived between AD 500 and 700. The Angles, Saxons and Jutes, who came to East Anglia in search of a better life, would in all probability have approached Ashwellthorpe (then unnamed) from one of the tributaries of the Waveney river, which then reached nearly to the sea in the south. The dark ash wood would have risen from the lush marshlands and mud banks, a welcome sight on the terraqueous landscape, providing shelter and respite from the sun. Anglo-Saxons tended to live in isolated farmsteads of one or two family groupings, eschewing the larger towns and fortresses founded by the Romans in favour of a simple rural life.

Lower Ashwellthorpe Wood has an industrial history that dates back centuries, serving the local communities with faggots, hurdles and poles. We pass a group of ash that has grown in straight lines and would make strong, flexible poles, essential if you were going to build a structure of any kind. The *Farming Journal of Randall Burroughs* provides evidence of the various uses made of Ashwellthorpe Wood. Burroughs was a gentleman farmer from Wymondham who maintained a log of his daily business that is of little literary interest but of great historical value. We can connect the ash trees of Ashwellthorpe Woods with Burroughs's regular acquisition of hurdles. On Monday, 21 February 1796 he states: 'Fetched a load of hurdles from Ashwellthorpe Wood for Wm Gray.' Gray was clearly a local veterinary surgeon, or someone who assumed the function of one, because on Sunday, 24 February 1799, Burroughs wrote:

> On Wednesday Nelson went to Ashwellthorpe Wood
> for a load of hurdles namely 3 dozen & deposited
> them in the 12 Acre. On Thursday to Wicklewood &
> borrow'd 3 doz gate hurdles of Mr Bernard. On
> Friday the ewes and lambs viz 46 ewes, 49 lambs & 21
> wethers were hurdled upon the 12 Acre. The lambs
> had been gelt on Tuesday by Wm Gray. The night was
> very rainy but they all recovered without ointment.

The hurdles he purchased were almost certainly made of ash and were used traditionally to hold livestock in place for routine operations, such as neutering or shearing. It is an example of how ash from these woods has served the needs of the farming community and advanced their interests.

Prior to the Norfolk Wildlife Trust's acquisition of the wood in the 1980s, it had been owned and worked by the Co-operative Society, which coppiced ash poles for broom handles and hazel brush for their heads. For centuries until the 1970s a brush factory in Wymondham obtained all its ash poles from Ashwellthorpe Woods. It employed many local women, who were highly skilled and worked for half the wage that men received. The Victorian historian William Kiddier recorded an ongoing battle that began in 1829 when male journeymen complained that the women were undercutting them. Kiddier laid the blame at the feet of the employers for exploiting the women, who were often so poor that they were simply glad of the work to support their families and did not dare to quibble over their pay. There is now a museum of broom-making at the former factory in Wymondham, which describes the plight of the local women who worked there for generations.

Lower Wood still has the feel of a working wood, with one well-trodden path in a regular shape, as if it was marked out by lines of coppicing and collection; there are few of the smaller tracks that children make in woodlands used for leisure. The clumps of thick, taller ashes seem impenetrable: the trees seem to have closed ranks, as if to prevent human foraging. Perhaps it is their response to the over-exploitation of centuries of industry, or maybe they are hiding the signs of Ash Dieback that are evident in their bare crowns. Steve Collin knows of trees in Norfolk that started showing signs of incremental growth loss six years ago but at the time no one realized what it was. That particular plantation has completely succumbed to the disease.

At Lower Ashwellthorpe Wood, there will be one final harvest of the timber that can be sold for firewood, when the dead trees are coppiced and removed in the cycle of woodland management. Once the infected ash has been coppiced it won't grow back, and infected timber can't be used because the disease stains it. Steve was advised by European experts to fell any ash the instant he saw wilting leaves, so that he could sell the wood before Dieback took its full effect. He wonders whether that was what happened on the Continent, whether foresters felled the trees with Ash Dieback at the first sign of the disease so that they could still make use of the wood. Such radical action would be detrimental to the ecosystem of a wood such as Ashwellthorpe, where removal of the dead wood from a forest floor with easily compacted soils and wet conditions would also take away the fungi and flora attached to it.

There are many schemes around the country, funded by the Forestry Commission, to burn wood as an alternative to fossil fuels but they were based on the reliability of ash, which stores well, grows fast and burns better than any other wood. The ash coppiced from Lower Ashwellthorpe Wood would have enabled the woodlands to pay for themselves. Now Norfolk Wildlife Trust will have to find other sources of fuel, which may mean buying in from abroad. This would not only be expensive but potentially fraught with hazards, such as the fear that the lethal

Emerald Ash Borer from America, where it has laid waste to millions of ash trees, could somehow find its way across the Atlantic.

We pause for a moment to watch a furniture beetle and a Minotaur beetle, distinguished by its horns, crawling towards a hole in an ash, as if in a race against one another to get inside it first. The hole is almost as big as the backside of a cow and the bark has peeled back to form a thick fringe framing the exposed wood, which is gradually rotting, aided by the beetles and other insects that feed off the bark. Steve picks off the furniture beetle to show me but it walks off the end of his thumb and falls into the grass: we have inadvertently helped the Minotaur beetle to win the race. Certain epiphytes, like lichen, prefer the ash bark because oak is too acidic and there is a danger that many such rare species will be lost unless they find alternative habitats.

Under the tree a mat of ground cover sends its leafy runners in all directions – it stretches right across the path. This is Ground Ivy, or 'ale hoof', as the early Saxon and Norse settlers called it. The leaves are very different from those of ivy, larger and rougher, with teeth all the way round. I rub one between my thumb and forefinger and smell it: it has a tart citrus flavour, a little like the lemon sage I grow in my garden. Apparently it was used as an alternative to hops for making beer, and the Scandinavian Vikings, who lived near the wood, would almost certainly have used it to make a fresh light ale. It likes damp places and semi-shade, so the foot of an ash is the ideal place for it; its purple flowers cling to the ivy, each with two lips to drink the rainwater. It is likely that when the Vikings dug it up they would have used spades with ash handles.

Before leaving the wood, I stand still to look up at the canopy of a few mature ashes growing so close together they probably share the same stool under the ground. Some of their branches interlock but their crowns fan out into finely etched sprays, each leaf like a black flower, hundreds of them dancing under the raindrops that tickle my skin. Speckled light filtering through the leaves almost makes me dizzy: the overlapping sections of the canopy seem to revolve, like a kaleidoscope turning. A nearby broadleaf tree, probably a sallow, forms a solid clump of closely knit leaves in its canopy; a morris dancer next to a ballerina, it is lacking in elegance and lets no light through.

The morning after the rain, the clouds gradually clear, giving way to blue sky and warm sun: a perfect English summer's day. My friend Lizzie and I walk away from the village towards Underhill, slightly uphill but not so that you'd notice. The farmer has cut a path that takes us straight across a corn field. Wheat ears brush against our legs then spring gracefully back into place. The earth is soft and sap rises through the grasses as the moisture of the night's drenching evaporates on the warm air. Miles of cereal fields are spread all around us, relieved only by hedgerows and distant houses. A skylark is singing above, and I turn to see the wood behind the village. The plain houses of Ashwellthorpe have receded from view and it is easier now to imagine the wood as it once was, surrounded by the fields and marshes of the Anglo-Saxon settlement.

On our approach to the solid cluster of houses that forms the hamlet of Fundenhall, the fields that scroll away from the woods, with the marks of previous inhabitants imprinted on the soil, are used to human feet. The early Anglo-Saxons, when they arrived from their cold wastelands, must have been seduced by the warm climate and fertile, waterlogged land. We are walking in an anti-clockwise direction: south-east to Fundenhall, north-east towards Toprow, north-west to Wreningham, then south towards Lower Wood. From every angle I can see the wood. It is the dominant landmark in the area, the nucleus around which everything else has grown – fields, hedges, paths, roads, farms and houses, people and animals. In a landscape where hills are absent, birds of prey perch on the highest trees (probably ash) to survey the surrounding countryside for voles and mice.

The wood is about a mile to the west as we walk by a high hedge at the side of another neat field of corn. By the end of this walk we will have seen it from south, east and north. Between us and the wood, hidden under layers of earth and crops, the Anglo-Saxon burial ground lies behind Ashwellthorpe Hall, invisible now among the tranquil expanse of fields. Another skylark is singing above us, and the scents of flowers are delicate and sweet. Yellow Field Pepperwort, red Sheep's Sorrel and Elderflower waft across the golden waves of wheat. My excitement grows at the prospect of walking in the wood again. It is hot now, well past midday, as we cross the corn fields towards Lower Wood. Although I was in it less than twenty-four hours ago, I am eager to see it in sunlight.

A row of ash and alder guard the north side. They have been here longer than any other species. We are entering by a path where tall grasses have been trodden down and the sides of an earth mound have been sculpted by the weight of many feet. Lizzie is slightly ahead and pauses to look around. In black T-shirt and shorts, she is framed by an ivy-smothered trunk and a mature ash, almost hidden under the burgeoning mass of vegetation rising above her. She looks slighter than usual and, fleetingly, as she slips into the waiting wood, she seems like a tree wraith or a wood nymph.

An ancient bank, with a ditch, surrounds the wood on all but the west side, where Lower Wood was separated from Upper. The ditch, now rank with stagnant water, would have been deepened by farmers over the centuries to deter livestock from entering the woods, but the original bank is believed to have been created in the Anglo-Saxon period. The ditch would have been dug with ash-handled spades, labour-intensive work. The farmers and peasants who toiled over it are most likely under the soil in the burial ground I mentioned.

When the ash trees have all gone, the mound will be exposed. Nothing is ever planted in this wood. The 'Ash' in Ashwellthorpe will be a historic reference. The 'well', originally 'weall', meaning 'bank or mound' in Anglo-Saxon, will remain: a lip of mud sculpted over fifteen hundred years. It will continue to accommodate the rain and silt and keep the boundary of field and tree. If the bank could talk, it could tell us who made it and why, tales of the many people and animals who have passed over it and left. When the ash trees have gone, the bank will be the sole keeper of the woods.

The earthworks will do the job they were originally intended for: to delineate the woods from the farms and maybe as a form of defence. Perhaps they were decorated with spikes or sharp implements to impale robbers and marauders crossing the woods to raid farmsteads at dead of night. Protection was much needed in the long and turbulent period of the many Viking invasions of the eighth and ninth centuries before the Danelaw was established. It seems likely that the Anglo-Saxons conceived the earthworks as part of a defence system, as they were in other parts of the country. It was common practice for earth mounds to be built around settlements as defence boundaries.

The copse is cool and welcoming after the heat of the open fields, the air heavily scented with the humus of rotting vegetation and moisture-locked soil. A little light filters though the fine canopy of ash and sallow, but not enough to dry out the wood after the drenching of the last few days. Rays of sunlight sparkle on the woodland floor. Last night the branches were dripping and bent under the weight of rain; today the boughs of the ash are still heavy and there is a sombre air among the trees. Our mood changes as we walk through them, looking for lesions on the bark of young ash. At first it is a game to try to spot the signs and I'm keen to show off my new knowledge. But Lizzie is walking fast ahead of me, and doesn't want to stay too long to examine the diseased trees. We become silent, and I feel as I do when visiting a sick relative or friend in hospital: I want to stay and cheer them up but feel helpless.

We leave the wood to the plaintive whistle of a chiff-chaff, the two notes, one higher than the other, seeming to call us back. As we stroll past the meadow towards the car, I'm aware of the trees whispering behind me. I turn back, but the sound fades. Yet as I drive away from the village, with the windows wound down, I hear it again, many voices muttering, not words or syllables, but music-al notes. It reminds me of the description of the spirit chorus that the soul seers claimed to hear in Montaillou, France, in the fourteenth century. Many occupants of this wood are passing into spirit form and, clamour as they may, nothing can save them.

In my rear-view mirror, the dark body of the wood recedes from view, like a rain cloud passing by. I notice more ash with exposed antler branches and think of the many that will fall victim and die on the roadsides over the coming years.

The Science behind
Ash Dieback

People up and down the country are becoming more aware of the plight of the ash and, in an area outside Norwich, scientists are working to halt the progress of *Chalara fraxinea*. The John Innes Centre was set up as a charity in 1910 by John Innes, a landowner and entrepreneur from London, and has since been established as a centre for plant science and microbiology of international repute funded by the Biotechnology and Biological Research Council. The original buildings date back to the 1960s, when the John Innes Centre moved to Norwich, and the newer buildings have clustered around them in regimented designs of varying styles. If I didn't know better, I might have thought I was on the edge of a housing estate in London's Peckham or Brixton. When I walk across the car park on the same day in June that I visit Ashwellthorpe, I see that the buildings are softened by grass and a few clumps of trees.

Dan MacLean greets me in the reception area and walks me through a pleasant outdoor garden to the laboratories. They are empty, and looking through the glass wall makes me feel as though I'm being shown an exhibit in a museum. I remember as a child going round a futuristic exhibition at London's Design Museum of domestic life in the twenty-first century. The laboratories are a bit like that: high metal contraptions and long melamine tables, all very clinical. In fact, they resemble characterless kitchens, which need a few dirty plates and a fruit bowl to make them real. I can see four, each visible to the rest through glass panes, lending them a competitive edge – I imagine young scientists coming to show each other how it's done. At the far end, there is a flat-fronted grey machine, called an athemizer, where samples are frozen so that they break down into tiny particles or strands that can be analysed. There are little bottles, too, covered with tin foil, filled with transparent brown liquids.

The John Innes Centre is at the forefront of the fight against Ash Dieback. Daniel MacLean works for the Sainsbury Laboratory as a bio-informatician. He analyses DNA sequences on a computer. Ash trees are famous for distorting the division between male and female because they can be hermaphrodite, and a few are. They are often thought to be wind-pollinated because they do not produce petals or sepals and the flowers appear before the leaf grows. Pollen is a kind of plant sperm that is often carried by bees or other insects to mix with female structures. When greatly magnified, electron micrograph scans of pollen grains show them to be like circular pumice stones with tiny holes. They are pale yellow in colour and cluster together. When a tree trembles, its pollen forms a visible haze. Unlike the female trees, males do not produce fruit or seeds. The flowers of female parts on ash trees are purple and grow into seeds attached to 'keys' – so-called because they resemble old-fashioned keys. They turn rapidly in the wind and are also known as 'spinners'. Like humans, each ash will be unique but will share common characteristics.

In the autumn of 2012, when Ash Dieback was first found in the wild in the UK, Dr Anne Edwards, who also works at the John Innes Centre, took a piece of wood from a sick ash tree in Ashwellthorpe Lower Wood to the centre. Other scientists there took a DNA sample from it and confirmed that it contained the pathogen *Chalara fraxinea*. A cross-section of wood with Ash Dieback shows what resemble ink blots on top of the concentric rings. Windblown spores of *Chalara fraxinea* infect leaves. The fungus grows down the leaf stem and into the core of the tree. Trees with *Chalara fraxinea* are more susceptible to other pests and pathogens, such as *Armillaria fungi* or honey fungus.

Research into Ash Dieback aims to find out how the pathogen is getting into the trees. The work has been accelerated by the use of crowd sourcing, which enables people from around the world to make contributions. Whereas before Dan would have had to prepare a paper and wait for a conference, through crowd sourcing, or publishing results straight away, he receives feedback almost immediately. He is helping to build a map of the areas in the ash genome – the complete set of genes present in each cell of an organism – where susceptible trees differ from those that are resistant. Dan is a member of the Nornex Consortium, headed by Professor Allan Downie, leading the investigation into Ash Dieback, made up of various partners, which include the Sainsbury Laboratory, the universities of Edinburgh, York, Exeter and Copenhagen, Forest Research, the Food and Research Agency, the Genome Analysis Centre, and the Forest and Landscape Institute, Norway. Nornex Group scientists are working with Danish scientists who identified the so-called Tree 35, which has low susceptibility. If they can identify the unique genetic features that reduce its chance of being infected by Ash Dieback, it will help them to breed an ash resistant to Ash Dieback in Britain. Perhaps the tree contains an enzyme that inhibits the disease. Maybe its bark is thicker. They are trying to find the answers to these questions because *Chalara fraxinea* is such a virulent fungus.

Identifying resistant trees could speed up the process of replacing those ashes that will probably die of disease. In their search for sources of resistance, the task of scientists at the John Innes Centre and Sainsbury's Laboratory is made harder by the fact that there is almost free movement of plants: our border controls expose our native plants to exotic diseases because security is less tight than it should be. Dan confirms that saplings are grown in Europe then brought to these islands, which makes it difficult to monitor their provenance.

Early in August, Dan and his colleagues release the ash game Fraxinus on Facebook. In its first six months it attracts an overwhelming number of players, who score points by putting together sequences of coloured leaves on their computer screens, matching them to genetic data that scientists working on *Chalara fraxinea* have found. Scientists may use the data the game produces to help analyse the susceptibility of a certain tree to the disease or to probe genomic DNA.

At the end of August 2013, Antony Milek, a student, sets up a vigil to guard an ash just over his garden fence in Kitson Hill Road, Mirfield, in West Yorkshire. It is in danger of being felled because the rest of the mature trees close to it, nearly thirty altogether, have gone: workmen believed that permission had been granted for the land to be sold to developers. He attracts the attention of the local papers, who report on his activities. Antony sits in the shade of the tree, which has been there all his life, at the bottom of his garden: as he is so close to it, the workmen cannot risk felling the tree. The land in question was once a refuge for birds and small mammals. The ash towers over the fence; it has two trunks, and appears to be in perfect health. A protection order should have been placed on it because *Chalara fraxinea* threatens ash trees.

On 5 June in Arkon, Ohio, a woman is arrested for sitting in an old ash tree that developers are waiting to chop down. This is the culmination of a week-long protest by local people and supporters who have occupied the tree in an attempt to persuade the local council and the land owners not to fell it. The 'irony', the local paper reports, is that it is only a matter of time before the Emerald Ash Borer, which has already eaten billions of America's trees, will probably destroy this one too. Yet surely that provides a stronger argument to protect the tree for the duration of its life. The protesters are asking only to be allowed to enjoy the tree for as long as it lives, but the council and landowners have decided that, if its life is limited, they may as well remove it at their convenience. During the week of protest, the owners change their argument from strategic planning to health and safety: they claim that the tree's roots are raising a sidewalk, thereby presenting a potential hazard.

Financial gain is placed above human wellbeing. Clearly local people feel that the tree is an important focal point for their community. The furore surrounding its fate shows the intense connection that people feel for it and the stories that will be lost when the tree has gone. For a time it will be missed. Ash trees in Britain will be mourned, too, but let us hope that the John Innes Centre, and others like it, are successful in their endeavours to develop a variety of ash that is resistant to Dieback.

Secrets

In high spring Colt Park Wood still has a wintry aspect. It is a ghost wood in appearance and history, since it is all that remains of a much larger ashwood that originally enveloped the foot of Ingleborough, one of the Yorkshire Three Peaks, in the north of the county. It is believed to date back to prehistoric times, and is comprised mainly of ash, but includes a few other tree types too. Clinging to the lower north side of the peak, at a height of 350 metres above sea level, the trees form a long narrow strip of silver in the sunlight, the only visible woodland in the area. Their bark is bleached the palest grey, as if sucked dry of moisture. The spirits of the people and beasts who have lived in and around the wood linger here. Deforestation and its reduction to its current size has occurred gradually over the last five centuries, due to grazing by livestock, which eat seedlings and strip back the bark, and the felling of trees to clear space for fields.

It is the third week in May 2013 when I drive up the steep track from the Ribblesdale road to the Ingleborough National Nature Reserve with my friend Graham Mort, the poet. I doubt I would have found it if he hadn't known where to go. I knock on the door of the warden's barn, an enormous stone structure in a sea of meadow grasses. It is a dry and breezy day, good weather for looking round the wood, Colin Newlands tells me. He is the senior manager at Ingleborough National Nature Reserve so he knows a thing or two about Colt Park. The wood and the adjacent meadows and pastures, including Park Fell, are owned by Natural England. On wet days it is too dangerous to go into the woods because it is too slippery, Colin says. I walk through the meadows, which are dotted with the bright yellow heads of field buttercup and celandine, in sharp contrast to the grey trees. I follow the line of the fence until it stops at a wall where some rough steps, overarched with ashes, lead us down between two arms of the wood.

At the end of the path a flock of sheep scatters downhill. A bracelet of limestone embraces the terrace of wood, a cliff that in places reaches a height of four metres. In the rest of Yorkshire the ash has been in leaf for at least a week but, here, no leaves are in sight: the only green is on the occasional rowan sprouting from the rocks. One grows at a right angle out of the shallow rock face, its trunk as thick as a human thigh.

There is no entrance to the wood, as such, but there are a few places where you can gain a foothold and lever yourself up to the trees. Gaining access to it is not easy for people or animals, because when the Nature Conservancy, a predecessor of Natural England, took it over in 1962, it fenced the wood to preserve it as a sanctuary for rare flora and fauna. Colt Park Wood is shielded from sheep by the natural cliff face on the eastern boundary and dry-stone walls to the west where it joins the meadows and pastures.

I hoist myself up the rocks and crouch at the top to take in my surroundings. Lime-coloured lichen glows on every surface, giving the undergrowth an unnatural brightness. I stroke its lustrous coat, which has the texture of seaweed. It has spread its thick fronds over tree roots and rocks alike, muffling sound and disguising grikes, or fissures, which can be as much as three metres deep. Time seems suspended. Lambs bleat, admonished by the deeper voices of their parent ewes, but they seem far away.

The older trees are quite well spaced so plenty of light filters through the gaps and the rich flora can flourish. There are clusters of yellow primroses on the woodland floor, and galaxies of star-like white wood anemones. New ferns are curled tight, ready to open; the relatively rare limestone Polypody Fern grows here, its fine fronds waving above the rocks. When I lie on my belly and look down a grike, it is dark and fusty: the air of a tomb. There is humidity in the undergrowth: moisture retained in the earth beneath the stone.

Sheltered by the soft hump of Park Fell, the ashwood is a marvel of survival: its many common ashes are stunted like bonsai. Elsewhere, the common ash grows fast but on this open escarpment it does battle against the elements every day. Ashes often thrive on alkaline soil but, when measured, the trees show slow growth in girth and height in comparison with most ash trees. Average wind speed at ten metres from the ground is approximately 11 m.p.h. here, and a gale-force wind blows through on around ten days of the year. Consequently, the trees have worked hard to stay upright and grow. Their roots are sunk into the gaps between the rocks, hidden from view until you draw close enough to peer into the underground cavern beneath the limestone that supports the wood. Severe environmental conditions may have a detrimental effect on the trees' survival: the Colt Park tree survey, completed in 1989, shows that over thirty years, thirty trees had died. Such change within any woodland is normal, where storms, localized cankers and other bacteria can cause death in the occasional tree, but the arrival of Ash Dieback will devastate this ancient woodland.

The odd hazel or alder adds variety. It is known that the ash trees in Colt Park have a shorter life than common ash elsewhere in the UK, probably due to the adverse conditions of their habitat. Also, the life span of ashes in other areas is often extended by coppicing and pollarding. Like coppicing, pollarding is a pruning technique, but it is carried out higher up the tree, while coppicing occurs close to the ground.

Some of the trees are bent away from the wind, branches outstretched, as if to balance themselves. I step from stone to stone until, suddenly, my left leg slips down a grike. If I had twisted at the same time, I would almost certainly have broken it. This is no place for children or dogs, and you shouldn't venture here without telling someone: they can send out a search party if you don't return. Fortunately, I'm with my friend. Pausing to recover, I find myself at eye level with an Early Purple Orchid; at first glance I might take it for a bluebell but it is paler pink and each flower has delicate flaps of petals and a pointed spur. Its long leaves are covered with character-istic dark blotches. It is a rare flower that clearly thrives under the light canopy of ash.

The oldest trees are twisted, their long boughs, like winding tentacles, hovering protectively over the skinny saplings. Variegated mosses of mustard and lime hug the paving, softening the grikes and masking the deep clefts and sharp edges. From the corner of my eye, I fancy I see a holy man in a flowing robe bending to pick herbs. He bears a striking resemblance to Odin, with a long beard and unkempt hair. Odin, or Woden as the Anglo-Saxons knew him, had a habit of appearing among trees. In his book *Odin: The Origins, History and Evolution of the Norse God*, Jesse Harasta claims: 'He was a wanderer who appeared when least expected, bringing triumph or doom.'

The deep clefts between the rocks would be almost as effective as quicksand in killing unsuspecting walkers. Colin Newlands tells me that once at Scar Close, a nearby wood, he saw a red head lying at an odd angle on the woodland floor. For a moment he thought it was human but as he drew closer he discovered that a roe deer had fallen between the rocks and died.

Nearby I notice a makeshift wall of large limestone slabs. It would have been built centuries ago to keep live-stock out of danger. Some slabs have fallen away and the rest have slipped into a zigzag pattern, each layer at an angle to the one below, so that they form a line of marching stone gnomes. The militaristic stance of the stone soldiers reminds me of the many armies that must have passed through or near to the wood: the Viking army, which overwintered near here during the late tenth century, the Scots raiders, during the Scottish Wars of Independence from 1286 to 1328, and Henry VIII's army, sent to destroy Furness Abbey during the Dissolution of the Monasteries in 1537.

Colt Park Wood would have made the perfect trap for disabling the warring Scots who came to plunder Yorkshire villages and farmsteads. There are tales of Dalesmen building rows of cairns to ward off the invaders, but the ashwood would have provided the invaders with shelter. Perhaps some wouldn't have come out alive, though, thereby buying time for villagers to pack up and flee.

In 1950 the Ministry of Agriculture, Fisheries and Food had classed Colt Park as an area of good grazing for a variety of livestock, so the animals were free to roam, and until 1962 the southern end of the wood was a working part of the Colt Park Estate. Sheep have many reasons to be grateful to the ash, were they capable of gratitude. They huddle under the limestone bracelet when the wind is wuthering, because they know the top end of the field is warmer near the trees. The trees provided shelter from rain, ice and snow, protecting them from blizzards and extremes of temperature; in summer they offered relief from the sun's heat. The middle of the wood was finally closed to sheep in 1964, to preserve delicate and rare flora. The margins were mown for hay, which might explain why the mosses give way here to tough grasses. Generations of sheep from the same flock are shown their own grazing ground by their mothers and always return to the same spots. The grikes are not as frequent or as wide here, and there are fewer trees, probably because the sheep chewed them to the level of the pavement.

A large slab sits upright on the edge of the wood. It has an oval top in the shape of a rough-hewn headstone and to me it is a memorial to all the human and animal lives that may have been lost in the treacherous stone traps. I sit on a rock nearby to contemplate what might be found if the ash trees of Colt Park die. The fetid space beneath the suspended pavements contains thousands of years of forgotten history: an ancient natural graveyard, perhaps, containing the bones of unsuspecting Scots clansmen who fell to their deaths all those years ago, and of the sheep, cattle, deer, pigs and rabbits, which foraged here. There may even be evidence of occupation by the aurochs, an ancestor of cattle, extinct since the seventeenth century.

In the huge barn where the wardens work, they have a small study centre with tables and chairs for school groups and carefully stored exhibits in plastic containers. Here, Colin unpacks the femur bones of an aurochs, which were found in one of the area's many pot holes. On the lid of the box a diagram illustrates the size of an aurochs next to a human male: it stood six feet tall, like the man, while today's cow reaches the man's shoulder. Scoured with chemicals and light to hold, it is hard to imagine this bone supporting such a large beast. It would have taken courage to approach an animal of that size, yet the early Neolithic settlers domesticated some of them. They also hunted them, probably with spears made of ash, and may have built hurdles from ash to fence in those they kept for work. They would have harnessed the aurochs, like oxen, to cultivate the ground, using a simple plough made of ash. In 2011 scientists in Denmark collected DNA from samples such as this femur bone, then mixed it with some genes the aurochs shared with modern cattle, such as the Highland breed, attempting to resurrect the ancient beast.

On a damp day, I return to Colt Park with my family. We slip and slide down the wet stones on the green path to the bottom field. The dog and I skirt the limestone cliff while the rest of the family scramble up into the woods with their cameras, ignoring my warnings. The trees are taller, their fine fronded leaves resplendent in green. Last time I saw them the trees were grey and bare. Sheep are grazing, and one fat lamb appears to be chewing a twig, probably an ash windfall. The ash may have kept the sheep round here healthy for hundreds of years because the bark contains quinine.

I climb part way up the cliff and lean into the woods as far as I can: the delicate fragrance of white meadow-sweet drifts up to me. Once sprinkled on stone or earth floors to scent houses, it was also a sweet flavouring in mead or wine. I am startled by movement in the under-growth: something disturbs the grass and ripples a bush of the sweet wild raspberries that gleam like jewels. There are many clusters of the plants, heavy with fruit, ripe for the table.

We are climbing up a slope stretching down to the Ingleton road, which cuts through the valley. Beyond the road many gentle humps rise like sand dunes. They are glacial landforms called drumlins. Their names reveal much about local concerns: Hunter Hill, Goat Close Hill, Swinesett Hill and, higher up, Deer Bank. The original forest would have covered this field and the road, but was cleared before the 1600s, when Furness Abbey used the area for grazing and parkland. That was how it came by its present name: Colt Park. Colt Park Lodge was a stud farm belonging to Furness Abbey, which managed its land by dividing it into lodges or farmsteads. A high ground limestone pasture, such as this, contains more calcium in the soil than low-lying fields and would have made the perfect grazing area for young horses. It would also help to prevent them developing laminitis, an unpleasant disease that swells their hooves and leaves them lame.

We retrace our steps across the meadow, passing the barn and the farmhouse. Sheep graze on rough grass above Ribblehead Viaduct and a curlew calls from Park Fell above us. We are walking downhill, past the northern end of the wood, which I am sure would once have extended much further towards Gauber High Pasture, where we are heading. Above the quarry, a few piles of rubble are hidden among the grassy hillocks and sheep droppings. I climb up and walk around them, trying to make out the shape of the former structure.

As I stand facing Park Fell, the remains of the building are barely discernible, just tufts of grass and stones poking out of the uneven slope of the fell. When initial excavations of the site took place in the last century, it was thought that the structure resembled a Norse farmstead, and three Viking coins from the ninth century seemed to confirm this. The reason there are hardly any ruins left to study is not that archaeologists have taken them away: extraction of stone for sale became a lucrative business from the 1870s, reaching a peak in the 1930s, when a number of 'rockery merchants' sprang up in the area.

Whoever built here on Gauber High Pasture would have been well aware of the protection the trees would afford their families, livestock and crops in this unforgiving landscape. The trees create and maintain warmth, a serious consideration on a windswept hillside. It is likely that the wood stretched across the pasture, and that some of the trees were cut back for crops to be planted. The ashwood would have protected homes from heavy rain and strong wind and helped to keep the occupants warm, providing wood for fuel.

The Coucher Book of Furness Abbey, published in 1916, includes a reference to a hermitage near a wood on the land owned by William de Mowbray, who appears to have ceded certain land rights to Lord Adam de Staveley: 'and the hermitage will remain waste on condition that there will be none there except with the permission of William de Mowbrai save for the woods, meadow and pasture for Adam and his heirs'. Arthur Batty and Noel Crack, members of the Ingleborough Archaeological Group, developed a theory that the ruins of the building might be the hermitage, while the woods de Mowbray's document refers to are Colt Park ashwood and the pasture is Gauber. They reached this conclusion after spending many years searching the landscape and the literature relating to the area's religious houses; they located this spot on Gauber High Pasture following directions from *The Coucher Book.*

In 1974 when Alan King unearthed certain artefacts from the ruins of the larger section of the dwelling, he found a complete rotary quern for grinding cereals, a lath-turned spindle whorl for turning wool into yarn and a small bell. Batty and Crack concluded that these items, along with a sword beater of iron with a wooden handle, indicate that the inhabitants were highly skilled and self-sufficient. The presence of the ashwood supports their theory: it would have supplied wood for tool handles, herbs, some fruit and other flavourings for food.

It is possible to imagine a life for the hermits or monks, who would have walked daily up to the ashwood to forage, wearing their homespun robes. They would probably have had a sturdy loom made from ash. They would have grown their own crops, ground their grain with the quern and baked it into bread. These utilitarian objects have survived many hundreds of years to tell us about their simple daily lives. In Ireland, Wales, northern

England and Scotland, there were many such small monastic outposts on hillsides, beside lakes and rivers in uplands and isolated valleys in the early centuries of Christianity. The monks sought a life of solitude and prayer and were usually highly skilled in crafts, with a good knowledge of agriculture and herbs. Colt Park Wood would have provided for all their needs.

The demise of the ash trees of Colt Park is unthinkable, not just for the loss of habitat and shelter for flora and fauna, but because it is a place of magic and beauty. Such a microcosm is a rare and precious phenomenon in Britain today. When the ashes are gone the gales will howl across the fellside, with no trees to break their force. Denuded, the hillside will assume a bleak aspect that is less attractive to walkers. The flora and fauna that have thrived in the shelter of the trees for centuries will shrivel and die. Exposed to the lashing rain and wind, the lichen will wilt and the limestone become visible for the first time in a thousand years. The meadow flowers on the north side of the wood will last only a week or two in summer, exposed to the elements from all directions. Colin will oversee the planting of other broadleaf trees but they will not filter the light as the ash trees do.

Colt Park Wood will still be a Site of Special Scientific Interest and a national nature reserve but it may attract more archaeologists than conservationists. Concealed for centuries under the ash trees, the exposed limestone may become the focus of interest, and someone will undoubtedly apply for permission to dig between the clints, the slabs of limestone, and lift some of the slabs to unearth the history underneath. Alison Uttley is best known for her *Little Grey Rabbit* stories. She loved nature and appreciated trees, mentioning an ash in *The Country Child*, which was based loosely on her childhood in Derbyshire. In it, she writes: 'She covered it [a hole] with the stone, the secret the ash tree had always known.' The tree guards something hidden, as do the ashes at Colt Park.

Bread

High on a hill above Branscombe, in Devon, up a steep lane from the village, a grove of pollarded ashes huddle together, looking down over the lush fields of this fertile valley. Some are stunted by overuse and stifled by ivy; others are tall and ancient, strong trunks stretching high, their branches curling out towards the fields. Their roots must be deep because they cling to a slope of some 45 degrees. Those lining the lane reach across it to form an avenue of ash. Halfway up the slope, the trunk of one tree splits in two and the branches lean out to balance themselves. Dappled in sunlight, a stripe of bright grass grows in the middle of the lane, curving tantalizingly uphill.

One of the shorter trees is so tightly wound round with ivy that it is difficult to see at first what species it is. A few shoots have curled from its top and fine leaves are visible above the ivy, but the ivy is choking one of the more mature ash trees to death. Some of the newer branches are already dead. The thicker branches are bald, with mustard-coloured moss growing on them. The ivy reminds me of Porphyria's lover in Browning's poem:

> … and all her hair
> In one long yellow string I wound
> Three times her little throat around,
> And strangled her. No pain felt she …

The ash feels no pain and the ivy is not wrapping itself round its neck, because it doesn't have one. It merely shows that the tree is at risk, whether Ash Dieback infects it or not, and *Chalara fraxinea* is creeping insidiously towards Branscombe.

The ashes on the other side of the lane are tucked by the wayside in the dim shade of overhanging trees that climb up to meet the road above. Yellow tansy vies for space in the scrub with wild thyme, from which another ash protrudes. Once the focus of a thriving industry, these old pollarded ash are now neglected, although they are much in need of trimming. Peter Blyth, the head ranger of the National Trust in Branscombe who manages the trees, has been told not to cut any branches for fear that Ash

Dieback might enter the lopped branches. No longer working trees, free from the bonds of many years' service, they've put on girth and height, and each taken on their own individual appearance.

The wide canopy of a stately ash near the bottom of the slope flounces in the breeze. Pollarded frequently as a working tree, its trunk has split into seven and their many branches are splayed, like a Buddhist deity swaying her thousand arms in a dance of compassion. Sunlight sparkles on its leaves, forming a halo. It is the Queen of the Hillside, reigning over the sun-drenched valley. A tall thin one, wrapped entirely in ivy, resembles a large topiary hedge shaped like a giant chicken, a caricature beside the Queen.

A breeze hisses though the branches, cooling me as I sit on the slope beneath the ash grove. The village is busy today but no sound of voices rises up to this quiet hillside. It is just me and the trees. Suddenly, a lorry roars up the farm track, receding to a low drone on its downhill journey. I stand up and turn downhill too, away from the green haven, and walk to the village to meet some of Branscombe's elderly residents who will be able to tell me a thing or two about the trees.

In the valley, the sun bakes the lanes and fields, and the cows doze in the shade of the woods, ears twitching away the flies that plague them. If you listen, you can hear the swish of their tails and the hum of bees extracting pollen from wild flowers. In the same way, the village is a hive of activity. 'Branscombe' is derived from 'Branoc's coomb', meaning 'Raven's hollow', or from 'Bran', a legendary king in Celtic mythology. Branscombe is a steep valley with a river that drains into a small cove, Branscombe Mouth, where smuggling is supposed to have thrived along the rocky shoreline in past centuries. Always busy with tourists in summer, today, a hot Sunday in August 2013, there is a charity bazaar at the village hall.

I wander around it, skimming the stalls of books, crafts and cakes, looking for the bric-a-brac stall Maree Dowell said she would be managing. Maree has lived in Branscombe all her life. When I spoke to her on the telephone, she said that her husband, Ivor, knows about the former bakery and the ashes in Branscombe. Eventually, I find and greet her, but when she answers me I have difficulty hearing. I think it must be the cacophony of chatter in the hall and ask her to repeat what she said. The room seems to get quieter but it doesn't last for long, about thirty seconds. I decide that my hearing isn't as sharp as it once was, but I don't dwell on it. Maree takes me outside to introduce me to her husband and waves us to a table under a tree in front of the hall.

Ivor seems shy at first, unaccustomed to talking to people outside his circle, perhaps, or just naturally reserved. Either way, I decide not to switch on the recorder I've brought with me. He speaks with a Devon burr, which at times is quite hard for my northern ear to understand. He tells me a little about the fifty years he spent working on a fruit farm up the valley, picking fruit in season: plums, apples, pears and berries. Out of season, there was pruning and planting, deliveries and general farm work to be done. Yet one of the jobs he remembers most vividly was cutting back the ash and thorn hedges, helping to collect bundles of wood and putting them on carts to trundle down the valley to the bakery, where Gerald Collier and his brother Stuart would pay the farmers for the faggots – bundles of sticks tied together to use for starting fires. They burn hotter and faster than logs, and in Branscombe they fuelled the bakers' oven and the blacksmith's forge. The ash pollards on the hillside were one of the main sources of faggots.

Collecting enough faggots to keep the bakery ovens burning was a big business, and many people in the Branscombe valleys were engaged in cutting and collecting them. The trees on the hillside, now forlorn and neglected, provided many for the bakery, and they were pollarded on an annual rotation. Ivor remembers helping to take the bundles off the carts and piling them at the front and back of the bakery cottage. It took a few boys and men half a day to build piles as high as a house. Unlike a house, the wood pile was not hollow but dense and closely knit, the faggots heaped end to end, then cross-wise in neat rows, to take the weight of more without collapsing.

I walk across the road to visit the old bakery, in a yellow ochre cottage shaped in a large triangle with the low, sweeping thatch that is typical of the area. The bakery café offers cool relief from the heat on this early-August afternoon; its stone flags, worn away by feet over hundreds of years, retain the cold of the earth below. The oven is a wide slit, about a metre high and two metres wide, resembling a traditional Italian charcoal oven for baking pizza. It is lit by a candle that flickers over its cavernous interior where six rusty baking trays rest, aban-doned after the last firing of the oven in 1987. The bricks of the oval roof are airtight, built to withstand an approx-imate maximum heat of 230 degrees centigrade.

Guy, who runs the café, recalls that Gerald Collier used to come in every morning for coffee and chat after Guy opened in 1992. Guy had been warned that Gerald was a reserved man, but he told Guy all about the foibles of the building and shared many of the secrets of baking.

Even in retirement he couldn't stay away from the bakery because he had never known any other life.

Guy shows me a photo of the brothers. Taken in the 1970s, the two men stand side by side in front of the wall of ash faggots, which looms behind them, blocking out the light and lending a cold aspect to the scene. Neatly arranged in tight geometric lines, the wall resembles the stick houses built by nomadic tribes in the rainforest. The faggots must have been recently delivered because they are not covered with a tarpaulin: the bakery received two deliveries each year so they would have had to be kept dry.

On the ground around the pair are stray branches and loose bundles that they will collect and throw into the oven with the six faggots they put in every morning. I can almost feel the ache in their backs as they bend to collect the wood, as they have done every morning of their working lives, which spanned six decades. The sweet scent of ash smoke seems to emanate from the photo, and I fancy I can feel the heat of the bakery oven behind me, and smell the yeasty scent of dough.

Each faggot was made up of a generous number of ash sticks or large twigs, about a foot long and the girth of a man's arm. The faggots burned for about two hours until the ovens were hot and the wood had all disintegrated to hot ash. Then, with the oven at the maximum temperature, Gerald and Stuart would rake out the ashes and embers and wipe the oven clean with a wet cloth at the end of a long pole. This ensured that no ash stuck to the bread during baking. The doors might then be left open for a while to let the oven cool to the optimum temperature for baking. Mary Kenworthy, Stuart Collier's daughter, said: 'They never had thermometers. They could always tell the heat of the oven by the hot bricks.' The bakery itself became so hot during the firing of the ash faggots that the tins of bread could be left to prove on the long wooden tables there. In winter, when cold draughts swept under the door, the bread was left to rise in the smaller proving oven below the main one.

There are more photos of the brothers taken in the bakery in 1956 by a professional photographer, John Hucklebridge, who was documenting disappearing industries in the area. The interior is white and fresh, the bakers, younger and thinner than they were in the previous picture, smiling as they work. Gerald is unpacking rows of packaged flour lined up on the gleaming table, the same table that now stands as a scratched and worn exhibit. A tray of currant buns lies before him, and he keeps his head down as he works, his face registering his pleasure at demonstrating his profession. Stuart lifts a large baking tray ready to put in or take out of the oven.

The space is close and must have been roasting hot by the time the oven reached full blast. In another photo, taken at the same time, Stuart is alone, putting a tray of buns into the oven, sweat shining on his brow, his arms taut. It was hard physical labour, repetitive and intense, but ultimately satisfying, as the pride on his face attests.

Bread is part of our staple diet and the Bible, the basis of Western culture, refers to daily bread, meaning food in general. Gerald and Stuart would be artisan bakers today because the bread and cakes they made were a form of craft. We make do with shop-bought bread, often plastic-wrapped to keep it fresh, and feel lucky to buy bread made in a bakery. My family have always made their own bread. My mother was taught to bake it by her brother, also Gerald, the author, and she taught me. I kept it up when my own boys were tiny and we were on a tight budget, but once I started work again it became something I did to relax when I had time. Of course, the bread I make in a conventional domestic oven is nothing like the bread the Colliers made and delivered to Branscombe residents every day. Even the flour was local, ground at Manor Mill from corn grown on the fields nearby.

All the foresters and woodsmen I have met agree that ash is by far the best burner of all woods, though in London some bakers preferred gorse or hornbeam faggots, claiming they produced the best bread. According to Ivor Dowell, the faggots were cut from hedges and trees alike, so it is possible that gorse was one of the woods used in the faggots, but it was the ash that kept the bakery fires burning hot. 'The Firewood Poem' by Lady Celia Congreve was first published, apparently, in *The Times* on 2 March 1930.

Beechwood fires are bright and clear
If the logs are kept a year,
Chestnut's only good they say,
If for logs 'tis laid away.
Make a fire of Elder tree,
Death within your house will be;
But ash new or ash old,
Is fit for a queen with crown of gold.

Birch and fir logs burn too fast
Blaze up bright and do not last.
It is by the Irish said
Hawthorn bakes the sweetest bread.
Elm wood burns like churchyard mould,
E'en the very flames are cold
But ash green or ash brown
Is fit for a queen with golden crown.

Poplar gives a bitter smoke,
Fills your eyes and makes you choke,
Apple wood will scent your room,
Pear wood smells like flowers in bloom.
Oaken logs, if dry and old
Keep away the winter's cold
But ash wet or ash dry
A king shall warm his slippers by.

The use of trees as the main source of fuel, as firewood or wood converted to charcoal, decreased gradually during the twentieth century, so Branscombe represents one of the last bastions of a dying industry. As indicated by Dorothy Hartley in *Made in England*, coppicing and pollarding for industry began to decline towards the end of the nineteenth century, and continued to do so until 1970, which partly caused the collapse of its usual outlets. The need for firewood became almost non-existent as people preferred to use coke and coal, which were easier to come by. Coppicing was practised less and less in the north, and by the 1960s it is believed that there was active coppice only in Kent and Sussex.

On the steep slope where the Branscombe ash pollards are to be found, the woodsmen or farmers would have been in danger of falling into the valley. They probably used some form of billhook to cut suckers, or 'spring'. Hand saws and a branch trimmer or two would have completed their toolkit, with sturdy ladders, probably made from ash. The same skilled men pollarded the trees year after year, the skills passed down from father to son; the work required agility and an understanding of the land. The woodsmen would climb a tree with a billhook slung over a shoulder and work their way round it, lopping and removing branches. A couple of workers on the ground below collected and separated the branches into straight poles for manufacturing purposes or faggots, bundling the faggots as they went to speed up production.

The traditional time of year for coppicing and pollarding trees is early winter or early spring, depending on the local climate, which in Branscombe is mild with short winters. Both practices were carried out on the ashes on the hillside on an annual rotation, which served the needs of the local faggot industry. 'Cut not above half a foot from the ground, nay the closer, the better, and that to the south, slopewise.' This was the sound advice given by John Evelyn in his *Sylva or a Discourse of Forest Trees*, published in 1706 and quoted in a 1995 Forestry Commission pamphlet by Ralph Harmer. Harmer mentions that little has been written about the actual 'how to' of coppicing because it is a skill passed on from one woodsman to another.

Landscapes are constantly changing, sometimes indiscernibly to the untrained eye, at others dramatically, sometimes occurring naturally, at others caused by people. Nowadays most of us make our living from jobs that have little or no relationship with the land, and although Branscombe's populace was slower than most in moving away from more traditional subsistence labour, the majority here no longer work in agriculture. Ivor Dowell spent the last twelve years of his working life as a refuse collector because work on the fruit farm became too scarce. The ashes are a living monument to the time when the immediate environment served the working community and society of Branscombe by providing fuel for the bakery and the forge. Their demise is now inevitable. A living link in the chain leading to the past symbiosis of Branscombe and its people will be lost for ever.

Exposed to sunlight, the road and the grassy bank will dry out once the trees have fallen. Perhaps the trunks will be left where they fall, to rot and grow fungi or become insect palaces. Ten years later, only wild flowers will grow on the slope. Soil without trees to provide extra nutrients is not as rich and cannot produce the abundant variety of plants that the ash grove does. The bank and the road will remain, the farmhouse, too, and the cows and sheep will thrive. The giant hen will topple first, if it is not chopped down before it can do so. The Buddhist deity will outlive its mates – the ancient trees have more resistance to Dieback, having withstood many cankers and diseases over their lives, but eventually it, too, will succumb: its leaves and twigs will wither and its branches become brittle.

Ancestors

The roads that lead me to Keswick, in Cumbria, seem to be lined with ash trees, their yellow seed pods dancing over hedges that border fields. Sometimes a whole row appears at the edge of a wood. They occupy the liminal space between open field and high road: wind buffers, land markers, guardians of paths and highways. Threlkeld, meaning 'the slave's waterfall' in Old Norse, is one of the many Norse place names I pass, reminders of the Viking settlers recorded in the Domesday Book for whom the ash was sacred. The thralls, or slaves, would in all likelihood have been Welsh peasants. The Lake District has been so much visited over the years that Watendlath isn't as well hidden as I'd thought.

Two thousand years ago, the valleys would probably have been dense with all varieties of broadleaf trees. Now there are relatively few trees in the area, which is why the ancient ash of Watendlath, some more than four hundred years old, are so revered. It is late September 2013, early autumn on the fells, and some of the ash leaves have already turned golden or purple or even fallen. I'm on my way to meet Richard Richardson, a retired farmer. Until six years ago he managed Fold Head Farm, the only working farm left in Watendlath. Ill-health led him from the remote fells, where his was the only house for miles, to his bungalow in the centre of Keswick.

Richard has the tall, spare frame of a man accustomed to hard physical work and, as he gestures to his hearing aid and pats the seat next to him, I see resignation in his eyes. He gazes at a large framed photo of sheep being herded past Watendlath tarn and says, 'Ah, it's peaceful up there.'

As a boy of twelve he helped his father to pollard ashes in the field below Fold Edge Farm. 'I was just a nipper so they sent me up the tree with a billhook.' At that age, he could climb the trees with ease. A ladder was propped against a tree to be pollarded, and while his father used a hand saw to remove the thick lower branches, Richard would climb high to lop off the leafy thinner ones with the billhook. 'It was hard work,' he says, 'brought out a sweat on you.' He remembers watching as the leafy branches fell to the ground where the sheep were often waiting to nibble the bark and the juicy leaves. 'It were good for them,' he says. 'They couldn't get enough of it.'

Pollarding trees in those days was time-consuming: with just hand saws and billhooks it could take three men a whole day to do one tree. Richard took over pollarding the Watendlath ashes in the early 1960s when he acquired the tenancy of Fold Edge Farm. He continued to harvest tree fodder for his sheep well into the 1980s, but as his flock grew it became more economically viable to use hay fodder supplemented with molasses and other nutrients. He was farming through a period of enormous change in agriculture, when the pressure to mass-produce meat increased and subsidies to farmers were based on the numbers of sheep each kept. When he started at Fold Edge there were three working farms in Watendlath but now there is only one, which his son, Shaun, manages in much the same way that he did.

Watendlath was one of the last valleys to continue to feed the sheep over winter with its ash trees, but until the early twentieth century lines of lopped ash branches along the sides of hedges in fields were a common sight in the Lake District, left to feed the flocks over the winter months. This practice dated back to Neolithic times, and in the court rolls for the Manor of Windermere in 1416 the following rights were recorded: 'What Man or Woman dwelling within the Forest that felleth any wood for Fewell in the Wood assigned to his neighbour for Cropping but Ellers and Birks [other than alder and birch] forfeits 3s 4d.' In other words, certain trees, such as ash and elm, were reserved for 'cropping' to feed livestock, which was considered more important than using the wood for fuel. Wood was the most valuable of commodities in the fourteenth century for building, heating, cooking and furniture, and only fear of reprisal would prevent ordinary Lakeland people from taking a little ash or elm, the most useful timbers.

However, nobody could stop the sheep and cattle from helping themselves to bark, as John Clare's poem 'May' attests:

> Each hedge is loaded thick wi green
> And where the hedger late hath been
> Tender shoots begin to grow
> From the mossy stumps below
> While sheep and cow that teaze the grain
> Will nip them to the root again.

The Lake District is not the only mountainous area where tree fodder has traditionally been used to supplement the diet of livestock when hay is not available. In the mountains of Lacaune in the southern foothills of the Massif Central in France, the goats apparently can't get enough ash bark. In mountainous areas of Romania and the Basque Country in France, tree fodder is still cut annually and is particularly useful for holdings where the farmer has only a small number of animals. According to Ted Green, England's foremost ancient-tree conservationist, research has shown that 'The minerals, nutrients, micronutrients and trace elements trapped in the trees' leaves provide animals with elements that may never occur in present, modern-day diets.' It is known that the inner bark 'phloem' of ash contains water, protein, free amino acids, soluble sugars and starch, a percentage of carbon and nitrogen, macro- and micronutrients. In Scandinavia, ash trees have been used as fodder for livestock since time immemorial, a practice still continued today where ashes are to be found. In the Lake District, it is believed that large numbers of Viking settlers brought with them their cultural and farming practices from Iceland, Denmark and the rest of Scandinavia.

In February 2012, National Trust volunteers planted six hundred ashes, hoping to continue the tradition of ash in Watendlath. In the autumn of that year, Maurice Pankhurst, the forest ranger for the National Trust in Borrowdale, saw signs of *Chalara fraxinea*, which the Food and Environment Research Agency confirmed after testing. The saplings must have been exposed to the disease when they were planted in the Netherlands. A large group of volunteers and National Trust employees pulled up the saplings and burned them. Fortunately, most of the dead leaves had dropped into the tubing that supported the young trees, so the National Trust believes that the *Chalara* spores wouldn't have had time to sprout and be carried on the wind to the veteran ash in Watendlath. Only time will tell.

Hugh Walpole described Watendlath in his novel *Judith Paris*: 'Watendlath was an exceedingly remote little valley lying among the higher hills above Borrowdale. It could indeed scarcely be named a valley: rather it was a narrow strip of meadow and stream lying between the wooded hills, Armboth on the Grasmere side and King's How and Brund Fell on the other.' It has changed little. The geology and form of the land have remained relatively intact for thousands of years. Farmers, like the Richardsons, have worked the steep valley sides for centuries, moving their Herdwick sheep from one grazing site to another, fishing in the tarn and cutting their 'tree hay', a term coined by Ted Green to describe the use of branches for fodder. He hopes that farmers may return to the practice as a natural method of supplementing hay, which is so often damaged by rain.

In the summer of 2013 I go to Knepp Castle in Sussex, as part of my investigation into tree hay. It is owned and run as a rewilding project by Charles Burrell. On the day I arrive, in late July, it is so hot that even the birds seek the shade of the hedgerows. Ted and I climb into a buggy and set off. As we swing round the corner of a field we almost sink into the deep ruts formed by winter floods. Ted steers over the humps of hard clay. Clinging to the bars of the buggy, I brace myself for the bump but none comes: the vehicle was built for rough terrain. A welcome breeze brushes my hair and I take a deep breath, telling myself to sit back and enjoy the ride.

Suddenly we jerk to a halt beside a young ash in a hedge and Ted climbs out, walking round to the back of the buggy to retrieve his billhook. He reaches for one of the straight green shoots, pulling it towards him so that he doesn't fall into the ditch below. Slicing through a few, he hands them to me to place in the back of the buggy. The air is sweet with the sap of fresh ash, not as pervasive as the scent of new-cut hay, but aromatic as clover. We are collecting tree branches as fodder for the cattle on the Knepp Castle estate, which is providing alternatives to traditional grass hay.

The honey scent of ash sap hangs in the air as Ted starts the engine and we rattle over bumps and ruts, crossing a scrubby field. We're heading for Nancy Wood, the small plantation of mixed broadleaf trees that Charles Burrell planted for the birth of his daughter almost eighteen years ago, and the site where Ted and Charles began their tree-hay experiment. One of many schemes in their rewilding project, it has taken them a few years to get it off the ground, but the TV production team from *Tudor Monastery Farm* are visiting the week after, and Ted is making preparations.

At a young coppice of broadleaves in neat lines, we alight to look at the ashes that were harvested last year. The young trees are about six feet from the ground and appear too thin to be eighteen years old. Hidden behind the new shoots of green foliage, the amputees of last year's pollarding hold firm, waiting to sprout new leaves next year. Holly, elm and sallow stand companionably, most with a few central shoots cut off at between five and six feet from the ground. The cattle would browse most foliage of their own accord, but ash and elm appear to be favourites. Ted keeps last year's tree hay in a covered wooden barn and we stop there. It consists of branches with the leaves still attached and smells of humus or rotting leaves; it is made up of varieties of broadleaf trees, ash among them. Ted gets in and passes the branches out to me.

We drive round the fields and scrub a few times. Tansy and thistles are squashed under our wheels releasing their scent. We get excited when we pass a trail of horse dung that appears to lead downhill in the direction we're going, and veer to the right in the hope that we'll find the ponies sheltering in some woodland. All living things, except flies and bees, have taken cover from the sun this hot afternoon. As we slide down a track that looks familiar, I see another trail of horse manure and call to Ted. Then I notice that this is the same track as last time. The estate is divided into distinctive areas but today each wide hedge, stretch of scrub and patch of woodland seems similar, merging in a whirl of sunlight as we search for the Exmoor ponies Charles released here.

Finally, we spot a few cows munching grass near a shady coppice and the buggy grinds to a halt. Startled, they lumber up the slope into the safety of the trees, leaving one cow and her calf to stare suspiciously at us. Ted picks up his freshly cut ash branch and walks towards them, calling. I sit still, nervous of cattle. I have to admit that my fear is misplaced today because these are timid creatures. They seem not to recognize the branch as food, probably because it is not their usual feeding time. They have too much choice in summer. The calf bucks and trots into the woodland. The cow lumbers after it. Our chance of feeding them the ash branch recedes.

During the night, the deer come close to the front wall of the castle, and in the morning we find that the branches we discarded the previous day are now picked clean of leaves. Small-scale farmers in the mountains of Turkey, Romania and Italy still gather branches to feed their cattle. In northern climes tree hay was winter fodder but in Mediterranean countries it was used in the summer when everything had been burned off by the sun and it was the only 'green' feed available.

In Watendlath, climbing up the side of the beck after my visit to Richard Richardson, I hop over stones in my eagerness to reach the top of the valley. As I dodge the stream that trickles between limestone pavings, the misshapen angles of the stones caress my feet. It is an easy ascent, partially covered by trees on the wooded side of the valley beneath Grange Fell. As I climb higher and level with the ancient pollarded ash in the distance, they appear small and round, like clipped privet domes. The beck is fast here, gushing noisily down from the hamlet. I arrive just before the café closes, in time to sit with a cup of tea in the shadowy cottage garden.

The mossy tongues of lichen lick the damp benches and a fir tree obscures my view of an ash hanging over a barn. Its seed pods are browning and its leaves are yellowing. Rocky outcrops jut from the burnished golden heather on the fellside, and the colours graduate to the washed-out green of grass further down. In a 1914 painting by Dora Carrington, *Farm at Watendlath*, the fells rise so close around the white farmhouse that they appear to encroach on it, as if they are about to topple it, stifling people and sheep alike. The farmhouse is on Fold Edge Farm, where Richard Richardson lived and worked. On the side of the cottage a plaque reads 'Judith Paris lived here', for this humble stone cottage also features in Walpole's novel. Richard still returns two or three times a week, he says, just to hold open a gate for the sheep and chat to his son about the farm. He occasionally takes a look at the ashes.

As I walk back down the lane towards Ashness Wood, I pass a small field on my right where an odd-looking pair of ashes stands side by side. The shorter one was obviously pollarded last spring and its cropped head of leaves, carefully cut by Maurice Pankhurst of the National Trust, resembles an Afro. Cut down to six feet, its fat double trunk looks ridiculous next to its tall graceful brother. Its branches are strewn around it on the grass, the leaves and bark eaten by sheep. True to the valley's tradition, the National Trust has left the spoils to the livestock.

It is hard to pass the ash pollards without communing with them, so I climb over the low fence separating them from the lane and walk carefully, avoiding grass hummocks and rocks, towards them. Since Maurice Pankhurst introduced them to me last week, and I have been reading about them, they have come to seem like old friends. Each tree is an individual, entirely different from its neighbours. Many are twins, not birth twins but two trunks that have grown from the same root bole, possibly as a result of early coppicing, of being cut right down to the stool as a young tree. Usually one trunk is thinner than the other because it has grown from the fatter trunk, like the pair that form part of a short line of ash springing out of the field's rocky slope. As Richard Richardson told me, these are working trees: their limbs have been used over the centuries to make tools for local use, such as hay rakes or ladders. Their bark is scabby and stippled with lichen: the pH of ash bark is high, which lichen requires, especially in a moist northern climate such as this. Inside the hollow base of the tree a few curved stools are growing out of the tree. One is in the shape of a bird's head and beak, its wooden neck curled tight as if sleeping.

A short way up the hill a massive tree trunk, dark with moss, has been pollarded to prevent it tumbling down the slope. Its roots are thick and gnarled, the same colour as the rock that surrounds them. They have worked their way deep into the earth, and where they show above ground they have formed deep rivulets to drain the water, like exposed pipes. Another ash is embedded in the slimy stones and its hollowed-out trunk has spread so wide at the base that it appears to have two heavy feet, like those of a dinosaur, its legs bent at the knees. Next to it grows a thin hawthorn; a staff for the ash to lean on. One of the ash trunks has a peep-hole, perfect for an owl to perch in. Ted Green says: 'The boll or bollings on a tree are the cut surface, which hardens and can generate bumps and lumps.' This is consistent with the bark of these trees, which have been cut many times.

Sheltered below the knobbly rocks called Goats Crag, the ashes thrive in the high-sided valley. More grow by the beck on the other side of the road, green and healthy. Further down the road, a line of old trees staggers up the steep fellside below the overhanging rock of Reecastle Crag, at least two lying dead, gnawed by wind and livestock. A buzzard glides low over the trees and circles awhile, spying a mouse or a rabbit. Suddenly an army Chinook helicopter shakes the valley, its double rotors ruffling the tops of the trees. In its wake, heavy rainclouds pass overhead. I scurry under the cover of the woods and rush downhill to Ashness Farm, where I'm staying,

That night I reflect on the ash trees I've seen. I look out of my window in the bed-and-breakfast, and I can see the stars clearly because I'm in the middle of the country-side. We know that old ashes, of more than forty years, resist Ash Dieback for much longer than new trees, but we don't understand their link to the cosmos. In *The Man Who Planted Trees*, Jim Robbins suggests that trees may be able to bond with the stars in a way we know little about. He quotes Werner Heisenberg, a German physicist: 'What we observe is not nature itself, but nature that responds to our method of questioning.' Heisenberg implies that scientists don't know everything and there is more to trees than we can articulate.

I'm restless tonight. I've recently had some bad news that is hard to accept. I thought I had progressive hearing loss, but when I went to Harrogate Hospital for a hearing test, the consultant sent me for an MRI scan. I had to remove all my jewellery and personal items, and lie in a tube for half an hour. Within a few days I was called back and this time my husband, Mike, came with me.

We were ushered into the consultant's room and I knew from his expression that it was not good news. His voice seemed to be coming from far away. He said that the MRI scan had revealed a large growth on the right-hand side of my brain close to my ear. The good news was that it was benign but I needed an operation – soon. I would be transferred to Leeds General Infirmary to have it removed. The internet told me that only one person in a million is unfortunate enough to develop an acoustic neuroma, but when I told my friends about it a few knew of someone with the condition. All had lived to tell the tale, so I decided to trust to Fate and luck.

My diagnosis reminds me of the reaction that the news about *Chalara fraxinea* in this country prompted in the general public. Initially people were shocked and then it all made sense. Those in the know were not surprised: it was bound to reach Britain sooner or later. Plants and trees could travel freely around Europe and saplings seeded here were often grown elsewhere in Europe. When Steve Collin and Anne Edwards found evidence of the disease in Ashwellthorpe Lower Wood they already suspected what it might be; they needed only confirmation. Similarly, when I looked back I realized I had had symptoms I'd been ignoring. About a year before, I had woken a few times in the early morning feeling dizzy: the room spun and everything was out of focus. This engendered a fast heartbeat, through fear, so I saw the doctor to have my heart rate checked. The checks proved negative, though my doctor advised me to call the emergency services if it happened again. I remembered that in 2011 I had been to hear some music with my father. When a man had sung falsetto it must have affected my inner ear because I felt sick and faint. I had to sit outside until the concert finished. At the time, I got over it and didn't think any more about it but the warning signs had been there if I'd cared to put them together.

When I walk up the valley from Ashness Wood to Watendlath the next morning, the sun is rising above the crags, casting its watery rays over the red berries of the rowan and the yellowing leaves of the ash. Half the valley is still in shadow, emphasizing the contours of the hills and the sheer face of Reecastle Crag looming above the ash. Plump Herdwick sheep stand squarely on their chunky legs, oblivious of their fate. The gentle gush of the River Greta is the only sound to be heard above their blunt teeth cropping the grass of this green valley. Watendlath is two miles from the main road and in a century of its own. But things are changing, and in another ten years the ash trees that remain may be under threat. A government directive has decreed that all farmers who turn over a hundred acres of land to regenerating wood-lands will receive a stipend and, faced with the withdrawal of sheep subsidies, farmers are signing up to it, including the Richardsons of Fold Edge. The original intention was to reduce the number of sheep because soil erosion has been caused by heavy grazing. Trees will retain water in their roots and help prevent the land becoming waterlogged.

Around Watendlath Tarn there are more fat old ashes – and a few fishermen waiting patiently for a bite. Last year's pollards, their leaves flutter in the breeze and they lean their heavy weight on the uneven stone wall. One has a prime view of Fold Head Farm. The farm is not as neat as its whitewashed front suggests: it is rough-hewn, like the rocks around it.

Before I leave the tarn to begin the ascent to Stonethwaite, I pause to look back at the mist rising from the water. The sunlight has vanished and, if I didn't know it was morning, I might have thought it was dusk. I sniff the fishy vapour and the humus of rotting ferns, then spot the oldest ash I've ever seen, the grand old man of Watendlath. Richard Richardson told me about it. It is a hollow shell in a corner of a field that slopes down to the tarn and I can see the water through it. This is one of the pollards that have been 'cropped and lopped' since late medieval times. It is choked with ivy and a wild cherry tree has taken root on a branch. On either side of its trunk, above the browse line, there are two enormous curled horns, gnarled and knotted. Its branches twist and writhe like snakes, peering out of its leaves as if waiting to pounce. They remind me of images of the Nidhogg serpent that hid in Yggdrasil, the tree of life, the immense ash that connects the nine worlds in Norse cosmology. Like Yggdrasil, this ancient ash has its roots in water, and the murky tarn, with its vaporous haze, resembles Niflheim, where the spirits of dead ancestors dwell. Nidhogg means 'curse striker' in old Norse, and by nibbling at the roots of Yggdrasil, he intended to cause chaos in the cosmos, upsetting the status quo of the worlds balanced within the tree's roots and branches.

The Norse myths are currently enjoying a widespread revival in Iceland, thanks to Ásatrú, which the Icelandic state recognizes as a religion. At its heart are Yggdrasil and the Norse gods, Thor and Odin, whose stories have always been told to Icelandic children. Yggdrasil is central to Icelandic culture so it is no surprise that Ásatrú is so popular and membership of the Ásatrú Association is growing rapidly. Perhaps the Icelandic people are looking for true meaning in today's fast-moving world, and find it in the *Poetic Edda*, a collection of Old Norse anonymous poems, which tells of Odin, Thor and the worlds-tree, Yggdrasil.

The wind blusters, whipping my hair across my face, as I continue the slippery ascent above the tarn to Stonethwaite via Dock Tarn. Turning left, I read the National Trust sign requesting that I keep to the path 'to conserve the wildlife and to avoid damage to wet land'. There is no dry land, and even at the height of summer, I should imagine that the turf here is always soft with mois- ture, but I place my feet carefully on the slimy pavings and bend my head into the wind. On a stone shelf halfway up Watendlath Fell, I pause for breath and look back at the tarn and the hamlet. From this height and position, Fold Head Farm is a tiny nucleus of human activity, dwarfed on all sides by massive skirts of stone, just as Carrington depicted, that look as though they could crush it in an instant. The ashes dotted around the valley are like the 'warden' trees that were worshipped as protectors and bearers of good luck in areas of Germany and Scandinavia until well into the nineteenth century. Fred Hageneder, in *The Spirit of Trees*, writes, 'In Sweden, the Ash was one of the Vartrad, the guardian trees of the farm or estate.' They are the guardians of Watendlath, a self-contained mountain realm that has changed little since the Viking invaders arrived in the tenth century. They moored their longboats nearby at Seascale, meaning 'hut by the sea', and moved across the Lakeland terrain, so similar to the countries of Iceland and Denmark that they had left behind.

It was to Hugh Walpole that Nicholas Size dedicated *Shelagh of Eskdale*, written in the early 1930s, because the story was retold at his suggestion. The revival of Norse culture and heritage in the Lake District, which took place at the turn of the twentieth century and lasted until well into the thirties, was based largely on the study of Icelandic sagas, primarily *Poetic Edda*, which celebrates the mythology of the Teutonic peoples.

> An ash I know, hight Yggdrasil,
> The mighty tree moist with white dews;
> Thence come the floods that fall adown;
> Evergreen o'ertops Urth's well this tree.

As I climb the steep fellside, resorting to all fours to keep my balance and make faster progress, it occurs to me that I haven't seen anyone since I left the fishermen at the tarn. Only I am stupid enough to climb Watendlath Fell on a wet misty morning. I know I ought to turn back, but once I've embarked on a venture, I have to complete it. The top of Watendlath Fell is like a moon crater, nothing but rock and sky, the wind howling round my hood. I am alone in space, a sensation that is both liberating and frightening, so cold that I shiver in my woolly hat and many layers of clothing. It is a place to come to terms with things.

Odin, one of the foremost gods of the Norsemen and the father of Thor, willingly gave an eye to gain greater wisdom, or to be able to read the Runes. He made a pact with the giant Mimir, who guarded the well at the base of the Yggdrasil. That, at any rate, is how the story goes, but I can't believe Odin gave his eye so readily. On pictures he is shown with one eye in the middle of his forehead. My operation will leave me deaf on my right side but no one will be able to tell as I will look the same. I know that the team of neurosurgeons at Leeds General Infirmary has a very good reputation, but it's impossible to contemplate such a serious operation without anxiety. Ash trees with *Chalara fraxinea* have no idea what's in store for them, because they are insentient, as far as we know. I have consciousness so I have some idea of what's in store for me and I would like to run away from the tumour, but there is nowhere to go because it is inside me.

I continue upwards. There is no going back now and I'm very afraid. There are rocks all around me and the mist is closing in. Across the small stretch of water I can just discern the fells, rising like monsters, almost black against the sky. I pause for a moment and have to take deep breaths of the cold air to calm myself. It is like Iceland here, or how I imagine it. It seems much later than it actually is: it is dark because the clouds are full of water vapour. I'm turning away to start climbing the rocks when I see someone coming down. I nearly shout for joy: there must be a way through if he is descending. He's friendly enough but keen to continue on his way. I continue on mine, much comforted. It is not long before I meet a couple, also coming down. They tell me the way to Rosthwaite, a village on the round walk that I am doing, through there then on to Stonethwaite and over High Crag back to Watendlath.

Once I'm beyond the rocks, it's much easier going: there's more grass and a normal path. The views are good from up here too, though I can no longer see Watendlath. As I get further from the Watendlath ashes I grow nervous again because they are a fixed point of reference in this unknown landscape. There are more people around, though, and it becomes quite sociable, stopping to talk about the walk and the weather. A nice man warns me about the climb to Rosthwaite and his wife says she won't come again: she always finds it exhausting.

It is a steep ascent through some woodland, so sheer that I have to take it in stages and eventually sit down. The trees seem to have formed one dark mass that is closing in on me and I can no longer distinguish the individual species. I feel dizzy and sick. The ground starts to sway and I have to put my head into my hands. The tumour is making itself known. I have reached an impasse and I look around for someone who will help. No one is about and I'm not carrying a mobile phone. For a while, I'm not sure what to do but eventually I force myself to persevere, this time going downhill on my bottom. The ashes at Watendlath are my only known landmark and I have to work my way back there ...

As I cross a bridge, a couple direct me to the Royal Oak Hotel in Rosthwaite. When I go in I see a free table. I take off some of my layers and carefully hang my clothes on the backs of chairs. My face and body tingle from the exertion and cold, so I order soup to warm me. As I eat it, a strange sense of well-being suffuses me; I am glad I overcame that difficulty. When I set off again, the sun is warming my back and I wear my coat tied round my waist as I skirt Stonethwaite. I feel lighter as I start the climb back to Watendlath. Little do I know it, but this is to be my last walk alone in the great outdoors. I had no idea that what would happen next would change my life for ever.

Meanwhile, a variety of broadleaf trees, including ash, has been planted at Watendlath by the Forestry Commission and Shaun Richardson, Richard's son, and fences keep out the sheep. Erosion is a real problem in the Lake District, where rainfall is higher than the national average. If the ashes survive *Chalara*, they will continue to be pollarded every eight to ten years so that they don't topple down the hillside in the wind. The National Trust should achieve their goal: they aim to increase the number of trees in Borrowdale by 2024, and the ashes may no longer stand alone as wind buffers in the fields. They will be surrounded by younger broadleaf trees with dense foliage and new shoots which will suck up the moisture that currently keeps the ashes waterlogged. If Maurice Pankhurst still has care of the old ashes, I'm sure they will be left alone in their fields, survivors of a killer disease. I cannot think of a better use for them than to feed a flock of Herdwick sheep since they, like the ash, link us to our Viking heritage and the way of life on these fells, beloved of farmer and visitor alike.

Ash Fires

The trees climb the hillside, like a tribe of spindly youths converging on their elders in wavy lines. There are hazel, oak, birch and cherry. The rowans are the princes of the tribe, bejewelled with berries of ruby and amber, dazzling among the greens and browns that clothe their peers. They gather around the ancients, at the crown of the hill, that preside over the serene Exe Valley in Devon. Bordering the River Exe to the east and the Bickleigh Castle Estate in the north, part of Byway Farm's swathe of hill and glade once belonged to the estate. Recorded in the assize rolls of 1306 as Atteweye, which is Middle English for 'by the road', the farmland has been worked by John Greenslade's family since 1916.

From a short distance, the woods look healthy and strong, skirting a wide area of scrubland where low-lying grasses and herbs release their fragrance into the balmy air of a summer evening. In the mature wood there is a naked oak, its branches thicker and curlier than those of the ashes that surround it, washed out against the rich greens of the living foliage. It has been stripped bare by greedy grey squirrels, which have a penchant for the leaves, bark and twigs of young oaks. This is secondary woodland, trees that have been planned and planted deliberately, as compared to the ancient woodland at the top of the hill. The soil is not as rich in nutrients and ground flora as that of the floor of an ancient wood since it was recently open-field grazed by animals.

A few piles of abandoned tubes give the impression that more tree planting is under way, but they have recently been removed from the trunks of the saplings they protected. On closer inspection, the signs of disturbed earth on the scrubland are evident. It looks as though an army of moles has been at work for a week, tunnelling and turning the earth up and down the hillside, leaving the tell-tale mounds in their wake. This scrubland has been cleared, much as the ancient woodland here was cleared by the Anglo-Saxons and their beasts, to make more grazing land or space for building and fields. It resembles the photo of the New Forest in Della Hooke's *Trees in Anglo-Saxon England*, which she uses to illustrate what wood pasture would have looked like in the Anglo-

Saxon period. Ironically, 'Bicca' or 'Becca' is Old English for 'pickaxe' or 'mattock', a tool suitable for pulling out tree roots and breaking heavy earth and clay. 'Leigh' means 'woodland' in Anglo-Saxon, so 'Bickleigh' can be interpreted as 'cleared woodland'. Sadly, this 'modern-day' denuded grassland was subjected to an enforced clearance: it is where Farmer John Greenslade's fifteen hundred ash saplings, ravaged by Ash Dieback, were rooted until only a few weeks ago. Now it is their crematorium.

John said that the trees were hard to uproot even with the small mechanical diggers that the Forestry Commission brought with them, but his team of eight completed their task in a day. The roots were deep, strong tentacles holding on to life. I found a West Saxon poem dating back to the ninth century in *Trees in Anglo-Saxon England*:

> The ash is extremely tall, precious to mankind,
> Strong on its base; it holds its ground as it should,
> Although many men attack it.

These lines seem as apt today as they must have been a thousand years ago. John's ashes put up a struggle against the mechanical diggers: stubborn to the last, they resisted even as they were dying. Ash wood was often used for weaponry, since it was so flexible and strong, so its associations with conflict are deep-rooted.

I am sitting in the spacious kitchen of John Greenslade's bungalow, quietly taking deep breaths and sipping a glass of cold water. I resist the urge to scratch where my T-shirt has stuck to me. The view from the window is downhill towards Bickleigh Castle and I am mesmerized by the green and pleasant land John has lived on all his life. He is an affable man with a weather-beaten face from the outdoor life he has lived. A regular voice of the community on Radio Devon, where he earned the nickname Farmer John, he is not fazed by my Dictaphone and says he is keen to share his experiences so that people know the truth.

I'm wilting, exhausted after an arduous afternoon: I've spent three hours driving around Devon in the blazing heat, even though Tiverton is only half an hour from where I'm staying in Sidmouth. John told me to turn right up the M5 to Bickleigh, but I looked on the map before I set off and saw that Bickleigh was near Plymouth. I turned left. When I got to Bickleigh I rang Farmer John, who told me there were two Bickleighs. I sped back up the motorway ...

It is the summer of 2013 and the ashwoods at Byway Farm are the first mature trees in the south-west to be affected by Ash Dieback, a serious loss because woodland is scarce in Devon and Cornwall. Exposed escarpments, where strong sea breezes blow, make it harder for trees to stand their ground. In the south-west, surrounded on three sides by sea, a great deal of woodland has been lost over the centuries to building boats and ships, and for agricultural purposes. Yet the woods of Byway Farm, largely planted over the last eighteen years, are for recreational and conservation purposes and are to be self-financing.

Farmer John's grandfather became the tenant of Byway Farm, renting the land from the Fursdon Estate. The family bought it in the 1950s, and in the 1970s acquired five lush water meadows by the River Exe, perfect for their large herd of cows reared for their meat. In the mid-eighties, disaster struck in the form of a virulent tuberculosis outbreak that lasted more than two years, and also affected the two farms adjacent to Byway Farm. Much of Farmer John's herd was wiped out and in 1994 he sold the surviving cattle. At that time, a number of schemes encouraged landowners to plant trees and he decided to make his interest in conservation his primary concern. Shattered by the loss of his herd, he threw himself into planting with a passion. By the end of 1994 he had eleven and a half acres of broadleaf trees. Farming is a tough livelihood and Farmer John is not only resilient: he is entrepreneurial, building up a network of conservation partners and maintaining his developing woodland with great care.

The infected saplings were planted in 1995 and 1996, but John does not know where they came from. He trusted his suppliers and, like most landowners at that time, he did not think to ask about the trees' provenance. It was an exciting time for him and his wife, Janette. They designed the woods with the Sylvan Trust, which was running a scheme to help landowners establish and maintain woodland. John had found a project that would enable him to bring more children into the countryside, which had always been one of his and Janette's principal aims. They oversaw the planting of oak, ash and cherry, with the occasional walnut and chestnut, in wavy lines rather than straight ones, since the land is on an incline, around footpaths that wound up, down and across the land, forming a maze of walkways that lead naturally to the established bluebell wood.

The tree planting took place over a few years, as the consignments of trees arrived from abroad. John told me how easily the ash established, self-seeding and settling into its new surroundings faster than the other species. He showed me photos of the young woodland, curving rows of ash climbing the hillside, with bluebells and other wildflowers at their feet. The year 2013 has seen a surge in the number of wild orchids, further evidence that Ash Dieback is not a threat to flowers. In fact, since the change in land use, the flowers he remembers from his childhood have gradually returned.

Over the last twenty years, the Greenslades have created seventy-two acres of woodland on their 172-acre farm, an important contribution in a sparsely wooded county, as Devon is. Every spring they lead bluebell walks through their woods, and the pleasure John gets from sharing it shines in his eyes. In early summer, following the deluge of rain that battered the bluebells and darkened the bark of the trees, he climbed on his quad bike to drive round his woodland. He went slowly, looking for signs of life in his overflowing ponds and admiring the butterflies on the buddleia. When he arrived at the stretch of young ash, he noticed wilting leaves hanging off the branches. He climbed off the quad bike to examine the six-year-old saplings more closely, noticing that many of the branches were brittle and drained of colour. He reported his findings to the Forestry Commission, who confirmed *Chalara fraxinea*.

Soon experts and scientists were trudging over Byway Farm, looking for samples and studying their findings.

Farmer John lent them a large area of his barn to set up a makeshift laboratory. He showed me the *Chalara* spores under the leaves at the foot of the ash trees that remain: there are many thousands, white and flimsy like tiny mushrooms, forming a polka-dot pattern on the grass. Spores are the fruiting bodies that carry the disease. They spread it by breeding on leaf litter. In the growing season they can multiply fast but lie dormant in winter.

John showed me the more mature ash trees that are newly exposed where they used to join onto the younger ones that have been destroyed. They are grey and balding, with a few dead seed pockets hanging dejectedly. They bend backwards towards the healthier greener trees behind, as if startled to find themselves on the edge, instead of in the middle, of the wood.

A pit thirteen metres long by six wide, deep enough to hold and burn hundreds of ash trees, had to be dug. On a newspaper photo, the flames swirl into the dusk to be swept across the Exe Valley. The figures of family and friends watching are etched against the grey sky, dwarfed by the scale of the operation. The young trees are sacrificial victims in the battle against the disease.

Farmer John describes the shame he felt at having to hang the red warning signs for Ash Dieback at every entrance to his property, announcing to the world that the farm is a diseased zone. One of the things that upset him most was that they had to turn away the last few school groups booked in to visit the woodlands that summer. His local community has been largely supportive but recently a fellow farmer, whom he has known for years, took him aside and accused him of bringing Ash Dieback to the south-west, saying he should have known the provenance of his ash saplings. But even larger organizations, such as the National Trust, didn't check the provenance of their trees. Those guilty are the importers who knew that the disease was rife in the countries they were buying from, and the government took no action until it was far too late.

A month or two later I caught up with Farmer John in a telephone call. He told me that the farm had become a field laboratory for a number of researchers and scientists. Experts from the Food and Environment Research Agency, based in York, came later in August to install two spore traps near to the woods. They put in battery-driven machines, each with a wheel to catch the flying spores, close to where the trees are dying and at the bottom of the hill. On behalf of FERA, John changes the battery every weekend. He also removes each wheel when it is full, packs it and sends it to the FERA laboratory in York. As autumn approaches and fewer spores are produced, the wheels are not as full. Sporulation has decreased and

FERA predicts that it will have stopped by October. When the wheels are empty, John can stop changing them until next June, when the whole process will begin again.

It is in August 2014 that I see John again. That afternoon, the air is clear after rain in the morning, heavily scented with grasses and wild flowers. My husband, Mike, is with me and we arrive exactly on time. It is a year since I was in this corner of Devon and the changes are obvious: Farmer John has recovered from the initial loss of his ashes and is more upbeat as he gets into his 4×4. For me, the changes are not so good. So much has happened to me that I am not the same person. The operations I have undergone have changed my whole outlook. I will never take my health for granted again.

Farmer John's springer spaniel, Bella, runs in front of us as we drive up the hill. Before us is the Exe Valley, with the characteristic yellows and pale greens of Devon's fields. I see the world in Technicolor now: the rich greens of the leaves are brighter than before, and even the common sparrow makes me blink. Farmer John is kindly driving us because I can't walk very well, and this is the first research I've been able to do for my book since the operations last November. I have spent the year learning to walk and talk again after the removal of the large tumour that nearly killed me. I entered into the battle, like the aesclings, or Norsemen, before me, and won. Della Hooke refers to the fact that the ash was often used to make weapons because it is very strong. In Anglo-Saxon the word 'aescoracu' means 'battle' and 'aesc' is often used to mean 'spear'; in Beowulf 'aescrof' means 'brave in battle'. My survival is nothing short of a miracle and I feel privileged to be here. According to Hooke in *Trees in Anglo-Saxon England*, 'In Britain the ash remained associated in folklore with rebirth and new life.' This is certainly the case for me as I have been reborn.

Farmer John drives us past ponds that are heavily silted at the bottom because the summer has been very dry so far. He has replaced his ash with cherry, lime, hornbeam, a few wild service and some whitebeam trees, and all are now hidden inside their tubing, row upon row. He has labelled each tube with the initials of the tree inside. When the sun shines, the tubing is pale beside the green of the grasses. The far grasses are tall and almost pink in the sunlight, and the furthest trees are dark and ominous. The cherry trees are peeping out of the tops of their tubing. They have been the fastest growing, but they are not as fast as ash. A party of school children with learning difficulties helped to plant the ten wild service trees; these used to be a sign of ancient woodland and John hopes they will be again, when he is long gone.

All the types of tree that Farmer John has planted are mentioned in the January 2014 report by the Joint Nature Conservation Committee, 'The Potential Ecological Impact of Ash Dieback': 'These species were chosen to cover a range of management objectives and as being likely to regenerate naturally or be planted by woodland managers because of their suitability to establish and grow on site types that support ash.' The Forestry Commission has advised John not to plant all the varieties together, because blight may affect chestnut and oak. No native tree is as easy to grow straight and tall as the ash. It appears even to be unaffected by grey squirrels, which is unusual. Charles E. Little's *The Dying of the Trees*, written in 1995, told us what was happening in America but most people ignored him. Now it is happening here too:

> The replacement trees, set out in rows, all the same
> size and species, are less able to resist the drought and
> bitter cold, adventitious pests and diseases, because
> they grow only in simplified *stands*, not in the
> vigorous, complex ecosystems that evolved naturally
> over eons.

Only half of Farmer John's trees are veterans, but Little has pointed out how, left to themselves, trees often evolve a way of looking after themselves and future generations. Farmer John had planted his ashes in rows of the same species. Fortunately, the Forestry Commission is wise to the diseases that may attack our trees, and is aware that they will probably be attacked by one disease or another in the future. We didn't know about diseases that lay dormant in trees or shrubs until it was too late, when they had already spread.

As we progress slowly uphill towards the woods, I see more mature trees with Ash Dieback. They have lots of bare branches and last year's keys are still clinging to them, grey and lifeless. The disease will take another year to show itself in the older ones, but there are already signs. The Forestry Commission asked John to put fluorescent sticky tape on the mature trees that seem resistant to the disease and he did so, bright rings that cannot be missed. But he thinks even these trees will succumb in time because *Chalara fraxinea* is a killer that eventually fells them all. The Forestry Commission advised John to take down all the trees around a diseased one and, if it grew new keys in the autumn, to breed from it. The diseased trees are an eyesore for John and it is hard to see them in this state. Sometimes the disease begins at the top of the tree and works its way down, leaving a trail of destruction in its wake: dead and brittle branches, no leaves and the diamond-shaped lesions on the bark.

Farms nearby are also infected by the disease: a neighbour cut down several hundred ash while they were still healthy and sold them. Trees on the edge of the estate have it near Bickleigh Castle. Fursdon Estate has it, as also Lee Cross Farm. At the moment there is no compulsion to burn the trees, but the farmers didn't want the disease to spread further so they destroyed them. The implication for the Forestry Commission is that *Chalara fraxinea* is indeed spread by spores travelling on the wind.

The 4×4 bumps down a small track, darker than the others. The trees are thin and young here, growing close together. They are dead or dying, infected ash. John stops the car and climbs out to show us a spore catcher. The box rests on a three-legged frame, like a model of an early flying machine invented by the Wright Brothers.

There was a form of Ash Dieback in the eighties. Some testing was done but the government decided that trenches could be dug to stop it spreading. They had no idea about the wind-borne spores. Now this form of Ash Dieback is virulent, killing nearly all the trees it infects.

When John shows us some spores, we understand. He gets them up on a twig by rummaging around in the long grass. They are easy enough to find but you have to know what you are looking for. He puts a few on his hands and they are no bigger than warts. One spore will sporulate hundreds of thousands of times. It works by attaching itself to the leaf or, in some cases, to the bark. Each spore turns brilliant white and bursts, releasing thousands more.

Eventually, we arrive at a clearing where grey logs are piled neatly, almost white in the sun, contrasting with the bright greens of the live trees around them. They look more like bones, knobbly and bent. Farmer John calls it 'timber that we can't get rid of' and laughs ironically. He can't sell these logs because they are dead ash trees, and the Forestry Commission has forbidden it. There are piles and piles of logs, with grass and weeds beginning to grow between them. The wood can be used only for making charcoal.

In *Made in England*, published in the 1930s, Dorothy Hartley writes: 'The great logs and lumps of charcoal shine like starlings' feathers, and crackle and sing as they cool.' Nowadays, we associate it with barbecues or soft drawing implements, but in the thirties it was used in aeroplanes, welding, oast houses, for drying hops, jewellery, gunpowder, biscuits, distilling and in blacksmiths' forges. The practice of burning wood to make charcoal dates from before the Middle Ages, and was in its heyday from the first to the third centuries in England, for iron smelting, at the beginning of the Roman period. Hartley suggests that 'almost as much charcoal may have been made in the days of chain mail and armour making'. No doubt they had spears made from ash.

A charcoal-maker, or charcoal-burner, was always in employment and he was often an itinerant worker, moving from wood to wood within a certain orbit. His profession was usually passed down to him from his father, as most traditional skills were. The quality of charcoal produced today isn't as good as it was when it was made in great metal vats relatively quickly in one day. The ash logs would make good charcoal but, having had specialists in to demonstrate, John has concluded that it wouldn't be cost effective.

John's employees cut down about a thousand ash trees in the winter. The worst infected were burned in February but the best timber was kept and put into the neat piles in front of us. It would keep someone in firewood for years to come. Instead of being burned, though, it will probably lie here, abandoned. Insects are even now crawling up and down the logs, occupying holes in the wood.

It is the summer of 2015 when I talk on the phone to Ian Brittain, a plant pathologist with FERA. He has been responsible for monitoring the spore-catching wheels that Farmer John sends to York. Scientists had found more spores in the wheel at the top of the hill, closer to the woodland. Ian tells me that the spores attach themselves to the ash leaves and, over the winter, become fruiting bodies. They tend to favour humid conditions, with little exposure to sunlight or clear skies. Wind-borne spores can blow a few hundred metres at a time, or more than twenty kilometres in a year. In Devon, Ash Dieback has spread in the wild, and there is evidence that it is killing hedgerows. From the twisted girdles of some of his ash trees, FERA suspect that the disease has been in John's trees for at least ten years.

In *The Greek Myths*, Robert Graves writes that ash trees are 'the power resident in water'. The River Dart joins the River Exe near to Farmer John's estate, and then flows past and onwards. There are many varieties of ash and they are ever changing throughout the world. The water is always there in its many guises, but the ash as we know it is going to die out. The trees represent the voices of the world: they are subject both to the diseases that ravage them and to the whims of people.

Life Cycle

Next time I visit the ash trees on the hill above Branscombe, in the summer of 2014, it is exactly a year since I was there. I climb out of the car on my stick and make my way slowly to them. We have parked nearby so that I don't have far to go and I see that the leaves are verdant and the branches sway in the breeze of an early August evening. I'm so pleased to see them again – they're like old friends. They are even more overgrown than before. Peter Blythe's instruction not to lop them has remained unchanged. Ted Green says that pollards 'sooner or later give in to the weight of their top-heavy limbs' and I hope these don't topple down the hillside, but I know that the National Trust would fell them if they thought the trees were a danger to the public.

In folklore, the fate of trees and people has often been entwined: either one or the other suffers first, then dies, usually the person. This shows the closer link with nature, something that this society has now lost, whether this is seen as a good or a bad thing. In *Trees of the British Isles in History and Legend* J. H. Wilks refers to the Reverend Gilbert White, the eighteenth-century naturalist and ornithologist:

> Ash trees thus used as surgeries were preserved with great care, for it was thought that if, during the lifetime of the patients, the trees were either felled or injured, either the affliction would return or the patient would die simultaneously with the tree. White thought the practice was a superstitious ceremony derived from the Saxons before their conversion to Christianity.

I believe that people looked for their own condition or sickness to be reflected in trees. For example, ashes with a hole in the trunk were thought to cure a child with rickets. The hole was closed up with mud or clay and the tree grew so that the split could hardly be seen. When Sir James Frazer wrote *The Golden Bough* in the late 1880s, he mentioned a Thomas Chillingworth, then thirty-four, who had been handed through an ash tree when he was a year old and was miraculously healed, also apparently of a rupture. Thereafter he guarded the ash zealously because it was believed that if the tree was damaged so was the person. The connection was deemed to last until the person died.

Over the centuries the ash has been credited with the power to heal and has been included in herbal books since medieval times. Incidentally, my uncle Gerald Wilkinson describes in his *Trees in the Wild* how they were used as a cure for warts. The sufferer chanted, 'Ashen tree, ashen tree, Pray buy these warts of me,' while they stuck a pin into their warts, then into the tree. Gerald added, 'It is probably as good a way of removing warts as any.' There is no doubt that 'The bark, leaves and seeds are generally diuretic, laxative, blood-cleansing and help to loosen the salts of uric acid', as Fred Hageneder claims in *The Spirit of Trees*. According to him, Hippocrates made cures for gout and rheumatism from ash during the fourth century BC.

Writing on 'Trees in Literature' in *Trees in Anglo-Saxon England*, Della Hooke remarks on the close association between people and trees and quotes from an early Anglo-Saxon poem, 'Solomon and Saturn' which translates as:

For a little while the leaves are green,
then afterwards they turn yellow
and fall to earth and pass away and become dust;
such then is the fall of those
who for a long while commit evil deeds.

Trees were thought to wither when evil people died. I don't recall committing a serious misdemeanour, but I have been dealt an awful blow. I woke up from drug-fuelled dreams in hospital to discover I had lost the ability to walk and my speech was incoherent. The right side of my face was frozen after the operations, so I was speaking with only the left side of my mouth. Over the next few months, I felt frustrated and belittled by what had happened to me. I became thinner because I was fed intravenously for nearly a month. The hair on the right side of my head had been shaved off. My right balance nerve had been removed with my right cochlea, so I was wheelchair-bound. It was not until I arrived at a rehabilitation centre that I began to take my first steps with the help of a zimmer frame and kindly physiotherapists. Being unable to walk in the countryside was anathema, so coming back to see the ashes was special. Della Hooke quotes from an Anglo-Saxon poem called 'The Fates of Men':

> The beauty of the body is very fugitive and very frail,
> and truly like the flowers of the earth.

This seems very poignant to me.

I feel an affinity with the ashes: they look like I feel. Some are sick but others are growing leaves, showing they intend to live on, for now at least.

When I was in the rehabilitation centre, there was a small quadrangle that never got the sun, at least in winter. I loved the times when the physiotherapists took me outside the building as part of a session because it allowed me to get my bearings. My dreams consoled me: I dreamed of walking through fields and woods, under trees, ashes in particular. Interestingly, it was baking bread that helped me to recover when I was back at home. I'm sure the bread I baked was awful but my family dutifully ate it. The action of kneading with my right hand was good for me and it made me stronger and more confident. Bread is the staff of life, and creating it is satisfying. It makes you feel closer to the life cycle.

Earlier I referred to the ash as the tree of life in Norse mythology, Yggdrasil. In *The Spirit of Trees*, Hageneder suggests that it was originally a yew that, over the centuries, became an ash. In an October 2012 article on the BBC News, Science and Environment website, Jeremy Cooke had interviewed Moreton Kyelmann, a guide at the Danish Museum of Hunting and Forestry. He said: 'In Old Norse mythology, the tree of life is the most important thing of all. It connected everything. The legend is that when the ash tree dies, the world will fall as we know it. It will be the end of the Earth.' He didn't tell pre-schoolers the full story, he added, because they would be traumatized.

In the same way, I feel I can't tell everyone what happened to me because it is too much for many to handle. I had a craniectomy two days after the first operation to save my life. My brain was bleeding so a small piece of skull was removed to enable the blood to drain. I might have ended up with locked-in syndrome, meaning I would have been able to think and understand but not to communicate, or the operation might have failed altogether and I would have died. My family were facing terrible odds – but I survived. I have been very fortunate, although I believe that I held on to life because I was not ready to die. The *Poetic Edda* refers to 'Lif and Lifthrasir' meaning 'life and longing for life', which is all that I know and love.

Spring

In darkness, Wenlock Edge is a brooding presence, looming above trees and fields. It is the highest feature in the landscape but it is horizontal, with ash trees stretching against the sky as far as the eye can see. A sleeping black dinosaur that towers over fields and trees. The stars twinkle more brightly above it than if you face the other way. This is my first night in Shropshire; it's 2015, and we have walked to Wenlock Edge to see the woods at night. The tallest are ash, the predominant species in the woodland. The trees are stark against the sky, their branches curling. They are almost black and seem to be breathing freely now that few people are about. It should be silent but it isn't: the trees are rustling and an owl hoots at intervals. A bat flits across my path, probably looking for insects. It is twilight, the hour when magic is supposed to occur and witches appear. I have the feeling that we're being watched, as though the ashes have eyes or something is lurking behind them.

Magic is associated with Wenlock Edge, especially in the writing of Mary Webb, who spent her formative years within sight of it. She was familiar with the ashes because they have always been here, since time immemorial; they would have been part of her landscape. The Wenlock Edge of my imagination is intrinsically linked with the writing of Mary Webb, the England that we have lost, the way of life in rural communities that is gone for ever. Travelling into Shropshire, it gradually dawned on me that we were leaving the rest of the world behind. Life here is slower and I feel as though I've entered a time warp. A century ago, Mary Webb called it 'a Rip van Winkle of a place', and observed that 'Somewhere in the Middle Ages it had fallen asleep.' It seems to have woken up more slowly than the rest of England.

We were tired and arguing about the route. As usual, Mike was driving and, supposedly, I was navigating. Daffodils nodded their yellow heads at us from the hedgerows in the villages we passed. When we saw the trees of Wenlock Edge appear above the houses as we came out of Much Wenlock, we were silenced. The woods I had dreamed about were in front of me. At first only a few trees were visible but soon I saw more and more, which formed an unbroken line. The ashes, with their black buds, were easy to spot: there were so many. I wound down my window to smell the air and noticed a new scent: it had been raining and the trees emanated a particular fragrance of wood and earth. I couldn't sit still because I wanted to be roaming among them, looking up at them and breathing in the perfume of ash trees.

Our caravan site is below Wenlock Edge; we have to travel through it to get here. The trees are touching as we pass on the road beneath them, the ashes distinctive with their height. It is late afternoon by the time we unload our car and settle in for the night, but while we walk back and forth I am aware of Wenlock Edge nearby. I keep going to the door to look out or popping outside to breathe in the air because it is the scent of Wenlock Edge. Like Amber, the protagonist of Mary Webb's *The House in Dormer Forest*, I want to escape to the woods. In the novel, she refers to them as the Upper Woods, but I think it must be Wenlock Edge. The ashes, although taken for granted and barely noticed by most people, are synonymous with Wenlock Edge. Mary Webb tells us that every spring Amber waited to see the first blackbird singing in an ash. I have never lived here but I feel the same anticipation. That is why we end up on Wenlock Edge at twilight.

Mary Webb spent her formative years at the Grange, a low old house that is visible from the road near Much Wenlock. In *Daughters and Lovers*, a biography of Mary Webb, Michele Aina Barale writes, "'Nature offers the female adolescent refuge in a landscape in which", suggests Simone de Beauvoir, "she can experience herself as subject: ... She finds in the secret places of the forest a reflection of the solitude of her soul and transcendence ..."' This implies that Webb found herself in nature, as I did. She walked the lanes and paths of south Shropshire, and experienced a freedom that would be unusual nowadays, because of the number of cars on the road and the perceived fear of stranger danger. This informed her writing, and her oneness with nature is evident in all her novels; she could assume the speech of rural characters and knew all the local customs, cures and magic. This seemed to coalesce in her final novel, *Precious Bane*, about Prue Sarn, a girl with a hare lip who falls in love with a weaver. The title is from Milton's *Paradise Lost* and refers to the high regard for money, which, as Prue realizes, blights lives and love. Her hare lip makes Prue more conscious of her spiritual strength and the title acknowledges this too. The novel is written in the first person, the voice of Prue, and adopts her style of speech with its lyricism and the richness of her vocabulary.

When I graduated from university in my early twenties, my grandmother gave me a book that had been hers when she was young: an early edition of *Precious Bane*. It had a lasting effect on me: I read it when I was just discovering myself and my identity. Although set far from the metropolis, the ache of modernism throbs in Webb's novels. As a child I holidayed with my family near the Welsh borders, so I could imagine the countryside she described.

At the beginning of *Precious Bane*, Prue is an old woman recounting her youth, the events that led to her brother's downfall, and to her marriage:

He was ever a laugher, was the woodpecker, and a right merry laugher too. He'll fly into an ellum tree, and laugh to see it so green. And he'll fly into an ash, and laugh to see it so bare, with only the black buds and no leaves. And then he'll fly into an oak, and laugh fit to burst to see the young brown leaves. Ah, the woodpecker's a good laugher, and the laughter's sweet as a sound nut. If we can laugh so at the end of long living, we've not lived in vain.

It must have been spring, if the ash trees still had no leaves and only their black buds were visible.

It is just after Easter on the first morning when we turn left along Wenlock Edge and climb beneath the trees, shuddering with cold despite the warmth in the sun. I put on my jumper, which I'd tied round my waist. The whole family have come to Wenlock Edge – we and our two sons – so we're very happy. There are lots of ashes on our route. They like the soil here and are self-seeded. The National Trust, which owns more than eight miles of Wenlock Edge, never has to import saplings because they grow here naturally. They provide dappled shade for us to rest in. We sit on a steep bank and, far below, see a tractor at work in a field. A few buzzards hover above, scouring the earth for mice.

It is because my mood is light that I can ascend for about two miles. We walk along the ridge of Wenlock Edge, which this morning seems very much alive. It is slow going for me and I start to perspire from the exertion. I have improved a lot but it hasn't been easy: it takes all my energy and determination to go upwards. I have lost a lot of fitness and in my hiking boots, which have such thick soles, I wobble a bit unless I'm on grass or mud.

It is hard work, for me, getting to Ippikin's Rock, a steep ascent between trees and bushes, but a sign announces it so we know we're nearly there. I cling to Mike's hand because I can't go so steeply without him. I have a stick but that isn't much use when the gradient is so precipitous. We have to climb the last part when we see a rock jutting out among the trees. It gets harder and harder. Our sons are already on the rock and there is a moment when I pause: going on is difficult but I can't go back either. I resort to all fours because that way there is no danger of falling. I've always been afraid of heights anyway. I sit down with a thump on Ippikin's Rock and gaze, mesmerized, at the view. Below us a sea of ashes rolls back and forth in the breeze, like the tide, and beyond to the west, hills rise as if they are made of dough. The ashes seem so close that I could almost reach out to touch the nearest ones, but if I did I would fall through the canopy. We are high up here and I can hear their branches rattling. The top of the rock is made up of small stones, but it is flat enough for us all to sit on.

It is difficult to believe that Ippikin's bones lie beneath us. We are sitting above the lair where Ippikin, a robber who lived in the thirteenth century, is supposed to have stored his booty. It is said that he was a knight who terrorized the people of the area with his gang of reprobates. One night, in a terrible storm that shook the ashes, a flash of lightning hit the mouth of the cave, which fell in, trapping Ippikin and his men. Ippikin still haunts this place, and if anyone dares to call, 'Ippikin, Ippikin, keep away with your long chin', he will appear and knock the person off the rock. A similar tale is told by an old man in *The Shropshire Haunts of Mary Webb*, by W. Byford-Jones, but he says that if people call his name, Ippikin will appear and 'slash out with a chain'. So at least two versions of the same story are known locally.

For a long time, I take in the vision below me, wondering what is missing. At last, I realise that there are no stone walls. In Yorkshire all fields are bordered by dry-stone walls, but here no hedges or trees soften the partitions between fields, blending in with the grass. Dark clouds are assembling overhead, pale and close to the land. Even the branches of the ashes are turning a darker shade and we get up to avoid the rain. We climb down and it begins to patter. We walk towards the road and Mike goes back to get the car.

Mary Webb suffered from Graves' disease, which afflicts the thyroid gland, but not much was known about it then. The symptoms were a faster metabolism, nervousness and gastric problems. The physical effects included bulging eyes and an enlarged thyroid, which embarrassed her to the extent that she kept away from society and couldn't even wander easily through the fields, which she was in the habit of doing. She nearly died of it, just as I almost did of my tumour, and there are other similarities between us. I became more reserved and for a long time I seemed to lose the capacity for joy. I found encouragement from the way that Mary Webb coped. In her biography of Webb, Gladys Mary Coles writes, 'If she became a semi-recluse, it was not because she turned against humanity but rather that she turned increasingly to nature with an ever deepening dependence …' She felt a keen sense of physical inferiority, which is exactly how I feel; as though everyone else is stronger than I am. For a while, I was forced to stay in hospital away from nature. It was some consolation to watch the fish in their pond in the quadrangle, and I used to leave my window open a crack so that I could hear the birds calling to each other.

Mary Webb published a book of essays, *The Spring of Joy: A Little Book of Healing*, about how she found a cure in nature. Later in life, she became a pantheist, meaning that she found God all around her in natural things. She called this 'home-brewed' but it was much more than that: Coles states that 'Nature was for her both refuge and church, solace and spring of renewal.' She found inspiration there, and I was able to do the same. Far from stopping me enjoying the countryside, my illness has enhanced it. When I am among nature, I make the most of it. I feel I've been given a second chance and I see everything with fresh eyes now. I am deprived of my liberty but that makes me more determined to get back what I can.

Sadly, Mary Webb died of Graves' disease and pernicious anaemia, aged only forty-six. She was dogged by ill-health for most of her life. She left behind a rich body of literature for which she will be remembered. I often reread her work and each time I find something new in it. I am able to relate my own illness and recovery to her life but, unlike her, I live on for now. Each new day is full of wonder and I am grateful for the love that my family gives me. Without it, I doubt I would be here.

On another day the birds are whistling as we walk up Roman Bank. I can hear the squeaky call of a buzzard overhead in the trees. The slope is steep and very long; everything seems much further than it used to. I use a hiking stick and lean heavily on it. Turning right, we come to a cottage above Lily Wood, and I spy an ancient ash in the front garden with three thick trunks sharing the same enormous bole. The girth is very large and the tree looks a couple of hundred years old. At some point in the past it has been a working tree as it was clearly once pollarded, which may be why it has lived so long. This tree has served people's need for fuel or to make things. Of its three trunks, the upright one, possibly the original one, is wound round with ivy, while the thick one on the right is leaning forwards almost at a right angle. It has a ladder propped against it, with some wooden poles. Perhaps the ladder has been put there to help a child climb the tree. It is dark in the shade of the tree. I want to sit on one of the boughs and rest awhile. The tree has an air of neglect, as though the owners lead busy lives. There is no doubt that it has been in the garden longer than they have been in the house.

I am about to knock on the door when something stops me. Everything is quiet so they probably aren't at home anyway. The tree must have seen many changes and must have been loved, or perhaps just tolerated. It has been allowed to grow and has not been touched for years; its branches reach out and hang in curls. I do not make the owners' acquaintance, so they remain a ghostly presence to me. Mary Webb wrote a poem called 'Presences':

> Sweetly in April I have heard them calling
> Where through black ash buds gleams the purple hill.

Beyond I can see ashes that have recently been coppiced – there are piles of logs and branches at their base. There are many plantations on Wenlock Edge: it is a way of maintaining the woods and making them pay for themselves. The National Trust grows ash here and sells the timber locally by the ton. The trees are regularly thinned out to help them grow better because ash seeks the light and young trees are often too close together. If left to themselves, the young ash may develop cankers and become very thin. Alistair Heath, the area ranger for the National Trust, has never seen any diseases in the ashes of Wenlock Edge, so they must be very strong. There would have been more wych elms once, before they were killed by Dutch elm disease in the 1970s, and ash was a tree that speedily took over because it grows fast and is very strong.

In May 2015, according to Peter Carty of the National Trust in south Shropshire, evidence of *Hymenoscyphus pseudoalbidus*, or Ash Dieback, has still not been found in the ashes of Wenlock Edge. There was what appeared to be an isolated instance of it in newly planted saplings south of Church Stretton in 2012, but an occurrence is never a one-off: *Chalara fraxinea* spores travel on the wind. Some must have survived, even if the saplings were killed straight away. Alistair Heath tells me that to try to pre-empt Ash Dieback the National Trust is felling healthy ash to clear about an acre, then planting other trees. This will enable them to see which species grow fastest and healthiest on the soil of Wenlock Edge. At the moment, beech and sycamore are to be sown, because they are well established in the area. Lime and cherry also grow well, so plans are afoot to plant them too.

We walk along Jack Mytton Way, named after a local Member of Parliament. Tall ashes are dotted about and some have purple buds, which resemble blackberries but they are lighter in colour and tighter.

The path narrows, becoming more overgrown with ferns and brambles, since the wood here is neglected. We pass through a gate, and the path skirts a meadow above the line of the trees. I am holding Mike's hand, walking between him and our son Dan, who is at least a head taller than both of us. Our other son, Joel, is walking behind us, taking a photo of us which he later sends. There is a line of trees on our right, dotted with ashes, and fields going gradually downhill to our left. I stop to admire some little cuckooflowers; the blossoms appear to be white but when I look more closely I see they are pale pink. Vapour rises from the grass, the sun is warm on my face and I am with the three people in my close family. The feeling of happiness is fleeting but I remember it.

When we sit on a log to rest in Stars Coppice we can see the view towards Church Stretton. The colours are a chalky palette, the sky is still slightly misty, and through the trees the fields are broken up so that the reds and greens seem to merge into one another. We decide to split up because it is too arduous for me to go back up the hill and I want to see more of the woods. There were times when I thought I would never see them again, let alone be able to walk on an incline, and I want it to last for ever. For a moment, I stand still and listen to the ash rustling in front of me, as if it is whispering to me. A grey squirrel hops across the tree canopy, so light the branches scarcely register it. My sons are walking away, one uphill after his father, the other down. I like the feeling of being alone with my thoughts, as if that is my natural state. I listen to their footsteps receding; it is a long time since I've been off the public highway in woodlands. I've become so used to being able only to use flat paths, where anyone can go, that I've forgotten what it's like to be away from civilization. I experience a flash of certainty that things will be easier from now on: I feel stronger, more able to get about and to speak to strangers. Aren't I in the woods? Didn't I get there with just a hiking stick?

After a short while, I shake myself out of my reverie and follow my son downhill, walking slowly under the ashes. I hold his hand to steady myself. Last year's leaves are underfoot as we walk down a steep embankment with moss on the right side. Trees are above us, some of them ash, enmeshed in places so that they form a trellis overhead, enabling us to see the sky chopped up, like pieces of a jigsaw. When I turn to look back I see that it is just like one of the holloways mentioned by Robert Macfarlane in *Holloway*. He tells us that 'The word comes from the Anglo-Saxon *hol weg*.' The path is worn by centuries of feet tramping over it and it is easy to imagine the charcoal-burners or woodcutters or journeymen taking a short cut through here. This is one of many ancient paths, or holloways, still in existence across Wenlock Edge. We arrive at a crossroads and pause for a moment. A sign reads 'Private Woodlands. No right of way' and on the other side another sign says exactly the same thing, so we have no choice but to go straight on. In this case, we take the steep trail downwards and I walk in the bed of a dry stream with rocks hewn out over the years by water, forming a narrow indent wide enough to walk in.

The next day it is late afternoon as we walk through Blakeway Coppice under the trees. We are on a well-worn road under the crown of ash, and in the undergrowth the seeds of wild garlic are getting ready to grow. Somewhere in the wood a woodpecker's beak is drilling into a tree. I've heard them lots of times but rarely seen one. If you look up you can see where the ashes have rushed towards the light at the roof of the wood. They have grown up quickly because they like the light of Wenlock Edge. The road is on a slope and above us is the side of the hill, but on its other side, through the trees, there is a quarry. In the latter part of the twentieth century much of the income in the area came from the excavation of limestone. The limestone escarpment upon which the wood has grown is of great geological interest: it is said that it formed more than 425 million years ago when Shropshire is supposed to have been south of the equator.

It is overcast by the time we're approaching Major's Leap but at the last minute the rain clouds drift away and the sun filters through the trees, lighting the forest floor. It is much further up than we anticipated and it is tiring for me to go so high. I have to keep stopping but my fatigue vanishes when I spot a gap in the trees with a view over the fields. It is hard to get to and I have to hold on to Mike to keep my balance. As I stare out at the hills beyond, all my cares disappear. Somewhere below is the crab tree that the Major is supposed to have clung to when he fell. Major Thomas Smallman's family owned Wilderhope Manor during the English Civil War of 1642 to 1651. He had to escape from Cromwell's men after he had taken back from them his possessions, which they had stolen. He galloped on his horse to these woods and jumped off a cliff to evade the Roundheads. He survived but his horse was killed.

We can see the Wrekin, shimmering blue in the distance, the hill A. E. Housman mentioned in 'A Shropshire Lad XXXI':

> On Wenlock Edge the wood's in trouble;
> His forest fleece the Wrekin heaves;
> The gale, it plies the saplings double,
> And thick on Severn snow the leaves …

The poem seems apt today but I doubt Housman had powers of divination. He refers to the wind but the Wrekin seems central to everything: 'To all friends around the Wrekin' is, according to Coles's biography of Mary Webb, *Flower of Light*, a traditional toast in Shropshire. 'It plies the saplings double' suggests the anxieties that trouble us, and 'Severn snow' refers, of course, to the river, which flows through the area on its way to Wales.

The right-hand side of my face tickles, as if leaves are brushing against my skin. I'm used to this sensation but it's annoying. I shiver as a cold draught crosses my back. A microlight passes overhead, its drone breaking into the sound of the trees creaking and the leaves rustling. It is time to go.

We walk back along a stone path that winds over the limestone edge, then down wooden steps. Built over the rocks, the steps lead past an ash with two long branches to right and left. It is growing at an odd 45-degree angle and is prevented from straightening by a tree that has wrapped its trunk around it, as if trying to choke it from behind. As we walk along the ridge, smoke from a farm fire rises upwards over the fields, merging with the blue of distant hills and with the lime green of closer ones.

Mike and I have come back on our own. It is a sunny day in early May 2015, and lambs are frolicking in the near fields. We can see Wenlock Edge from our bed-and-breakfast, and Brown Clee Hill looms behind it, the whole covered with a white haze as moisture rises from the wet grass. In a bookshop in Much Wenlock I came across a small book called *Shropshire Folklore Ghosts and Witchcraft* by Jean Hughes. It told a Shropshire story about Brown Clee Hill that illustrated the perceived power of ash trees. It concerns the young daughter, Pleasance, of a Maisie Bloomer, who lived in Cleobury Mortimer in the eighteenth century and is believed to have been a witch. Maisie is supposed to have saved the life of Simon Bache, a local man, by curing him of a fever. Harry Bache, Simon's son, was instrumental in this cure because he was Pleasance's boyfriend. At Easter one year Pleasance gave Harry an egg as a present and led him to the top of Brown Clee Hill where she bade him put the egg under a hollow ash tree. He left it there for three weeks, as he was told, then went back to retrieve it. When he returned Pleasance told him to put the egg under a hen. It hatched and the chick became known as 'The Witch's Cock' because it had outstanding vigour. Whether or not the story is true, it does show that people believed ash trees were special.

Shropshire is hedged in by counties that all have confirmed cases of *Chalara fraxinea* or *Hymenoscyphus pseudoalbidus*, and ash is a common tree in the Marches, the border area between England and Wales. In 2012 managed sites in Ludlow and Tenbury Wells were found to have *Chalara fraxinea*. It has already wrought havoc in much of the Midlands and it is only a matter of time before the fungus comes to Wenlock Edge. It may already be there – it is impossible to keep an eye on all the trees. Once Ash Dieback arrives, there will no longer be an unbroken line of ashes stretching along the limestone ridge from the Wrekin to Craven Arms. They will no longer spring out of the soil and wave their beautiful heads. The ash will become the stuff of legend. The old ash in the garden at Lily Wood will die. As A. E. Housman says, in 'On Wenlock Edge the wood's in trouble', the ashes will suffer until the gust blows over.

Isolation

The letter lies on my desk for weeks with the address written neatly, its secrets hidden. It arrives at the end of May 2015 when the leaves are new on the trees. Spring is my favourite time of year, when the air is heavily scented and the blossom is out. I am in hospital when the letter arrives and don't get to read it until June. Sickness strikes me down for months. My surgeon suggests that it may be because I had lots of surgery on my head. I am affected adversely by the insertion of a shunt, a small device that fits into the left-hand side of my skull and drains excess moisture. I have hydrocephalus, otherwise known as water on the brain.

The letter is about Rassal Ashwood in north-west Scotland, from Helen Murchison who used to run Lochcarron Heritage Centre. It whets my appetite and fuels the fire that is alight inside me. During the days and nights when I lie in bed at home, I dream of Rassal Ashwood, of the trees and their shadows. The letter tells me that the local people are surprised by the attention that Rassal Ashwood attracts from outsiders because ash trees grow like weeds in the area. They often have to remove the seedlings from their gardens and boughs of rotten ash sometimes fall on walkers' heads, making them a hazard.

It is an overcast day in the middle of September 2015 when I climb out of the car by Rassal Ashwood. Helen Murchison is waiting for us. It's a still day and within minutes we're completely covered with midges; they land everywhere, even on Helen's face, when I thought that people who lived nearby were immune to them. They tend to proliferate when there isn't any wind. We have to take refuge in our car where Helen and I settle in the back to talk.

Rassal Ashwood is supposedly the northernmost ashwood in Britain, in Wester Ross. On the Ordnance Survey map, it is one of a few patches of green besides the fir trees of Couldoran, a large white house that nestles among the dark trees across the road. The eye is drawn to the wood because it stands out in an otherwise bleak terrain. Coire Each is the mountain whose unforgiving face dwarfs Rassal, making it seem much smaller than it is. All signs of civilization stand out here: a car going past, a building or the telegraph poles. On a slope, the rain-water drains away, which stops the area becoming water-logged. The tree roots also absorb moisture. Helen tells me that the terrain of the wood is completely different from other places nearby, since a projection of limestone runs underneath it and the higher trees stand on it. The upper part, which I will never see, consists of deep grikes (fissures) that form limestone terraces up the hillside. The trees grow out of the gaps or on the limestone pavement and can be seen in relief climbing up the hill with large expanses of fertile soil between them.

The grass is rough and uneven, boggy in places. There are clouds of cream meadowsweet, and plenty of spear thistles with their long stems. After waving goodbye to Helen and donning midge protection, Mike and I step over a stream and finally arrive at the gate to the wood. There is a hand-written, wind-blown notice warning that cows are in the woods with calves. Bracken grows fast in these woods and the cows keep it under control. They have more right to be here than we do: they are probably descended from animals that have been grazing here for generations. Nevertheless, my heart sinks because I'm scared of cattle. Until 31 March 2014, Rassal Ashwood was a designated national nature reserve, affording it special status, but Scottish National Heritage delisted it. As a consequence, it appears neglected: only essential maintenance takes place.

My uncle Gerald Wilkinson mentions Rassal Ashwood in his *Trees in the Wild* and describes the seclusion of the wood, its relative lushness in this bleak environment. Perhaps he was in search of solitude then, and he certainly found it at Rassal. I experience the same quiet here that he must have enjoyed forty-two years ago: 'The wind is cut off and the atmosphere is warm and scented with meadowsweet – at least, it is in summer.' The high fences and the gate you must walk through to go in signal change from outside the wood's boundary. Even the air seems softer inside. Lichens, of which I see many varieties at Rassal, are known to favour humid conditions: perhaps that is why the temperature in here is slightly warmer.

Mike leaves me alone to explore the woods. The breeze blows gently, swaying the leaves and grasses, sending me a little off balance. It doesn't take much to do that nowadays so I hold on to an ash trunk to steady myself. It is little more than a sapling – my arm can embrace it easily. I am on the edge of the wood looking in, tantalized by the vision before me. I gaze at the trees. I have travelled half the length of Britain to see them and they are worth the journey.

On the outskirts of the wood, ash saplings have sprung out of the soil. Rowans are dotted about and the more mature ashes rise proudly above the other trees. Many of them have moss growing on their branches, verdant fronds that decorate them. I move forward slowly, afraid at every step that a cow might emerge from the undergrowth. The trouble is that I can no longer run away. No creature appears, though, and I walk some distance unmolested. It is so quiet that I can hear my breath.

Gerald was born in 1926, exactly forty years before me, but I hardly knew him. I admired him and he was unusual in that he defied convention by travelling around in his camper-van, exploring the woodlands of Britain, but I know him mostly through my mum's memories of him.

The picture Gerald includes of Rassal Ashwood shows a gnarled ash surrounded by ferns, much as it looks today. The large photo is shrouded in darkness since the canopy shades the ferns and grasses. The wood looks mysterious, as though tree spirits come out at night. It appears to have changed little since Gerald was here … Some trees will have rotted, others sprouted and grown up. In *The Signature of All Things*, a novel by Elizabeth Gilbert, the protagonist, Alma, is an expert on mosses. The book suggests that: 'To the unschooled human eye, moss did not seem to move at all.' Moss favours shady places, like Rassal Ashwood, where it is moist and the air is clean. Gerald also includes a small photograph of trees growing out of boulders, showing the wood in sunlight. The rocks are lit up and the contrast between darkness and light is pronounced. The leaves soften the trees and the grass looks as if it is on fire. In a third photo a single strawberry and its leaves are growing among the grasses. A bird probably ate it soon after.

There is a grassy path through the ferns, which are abundant and in places form large fans, reminiscent of the jungle rather than the Scottish Highlands. I am walking towards Coire Each, which looms above me, inaccessible. All around ashes are in varying stages of growth. I pass some saplings among yellow Common Ragwort, their branches waving in the faint breeze. I don't have to go far to see older trees, their curved trunks outlined against the sky, their branches spread out with lots of room to grow. They are no more than a hundred years old but they are already very tall, their uppermost boughs looking down on the surrounding countryside. Their leaves sweep the sky. No human has passed them for a long time, I am sure. People don't often come here and the chances are that the trees are not familiar with humans. Cows have nibbled at their bark but no one has interfered with them. Some branches have been blown off by the wind and now lie on the grass, left to rot or to be a home to beetles and other insects. They resemble antlers that have fallen off stags, disembodied, once erect and proud, now blanketed with lichen.

I see some hoof prints in the mud made quite recently by cows, or so I think. I look around fearfully. Cows could wander easily among the trees and might appear without warning. It is known that they are protective of their young and may attack if they are threatened. I feel help-less, as though I can easily be trampled by angry cattle. My fear is not rational, and the chances of me being hurt are slim, but for some time I am paralysed with dread.

The areas higher up the hill are impenetrable, so overgrown that even if I could walk better I wouldn't be able to go there. This is a strange comfort to me. A fence has been erected round the perimeter of the wood to keep out animals, especially rabbits and deer. Helen Murchison thinks that rabbits did the most damage, nibbling at the bark of the ashes and eventually causing them to die. Rabbits were introduced in the late nine-teenth century, quickly breeding and spreading. It is said that during the Second World War the shepherd employed here made a lot of money trapping rabbits to send south for food.

In his dendrochronological analysis of 1986, George Peterken, an eminent ecologist, found evidence of a tree that dated from 1684, but it is believed that Rassal Ashwood has existed much longer than that. A black-and-white photograph of Rassal, published by Peterken in *Natural Woodland*, shows sheep grazing. The photograph was taken in winter when the trees were bare, and I can see in it part of a low dry-stone wall between two trees. There is lichen on the bark. The book claims that when the wood was left to regenerate a copse of hazel, rowan and great willow formed.

In an area of the woodland, not too far in, there is a grove of silver-branched rowan with berries on it. I am conscious of an unearthly atmosphere, very still and otherworldly. The trees appear to be in rows here, as if they have been planted purposely. They are obviously quite young because they have grown up but not outwards. I am able to walk among the stands of rowan, looking down the slope. Sound seems amplified; my feet crunch over the grass and ferns. Rowan trees are special to me because we buried my sister's ashes beneath a rowan beside a waterfall in North Yorkshire. I go there regularly to see the tree grow and to commune with Jane.

I stand for a long time, looking at the trees and the mountains beyond. Time seems unreal here: it is as if this woodland exists in another dimension where things appear to stay the same and the only changes are trees falling or birds dying. A telegraph line runs through the middle of the wood, a reminder that the outside world encroaches on this one. Access to the heart of the wood is difficult; impossible for me. I am able to see it from the outside, tantalizing in its lack of availability. I doubt that a human has penetrated the dark core of the wood for years. When it gets Ash Dieback, the disease may remain unnoticed for a long time. If something happened to a person in there they might get lost and never be found.

The north-west corner of the woodland, Helen Murchison tells me, was once a burial ground for people who committed suicide. They could not be buried in hallowed ground or within sight of the sea. There were thought to be two burial grounds, according to the report of Tom Cooper, a community woodland manager; the second was adjacent to the first and was apparently used for mine workers. I would prefer to be buried in the natural environment rather than in consecrated ground, vying for space with everybody else. I can think of no better place to be laid to rest than under the ash trees in the shadow of Cnoc nam Broc. It would be a long way for my family to visit my grave but it would be an earthly paradise, used only by birds, squirrels, rabbits, deer and cows.

I have recently received a diagnosis that has made me think more about my mortality. The tumour has grown back and is nearly as big as it was before. It seems I have a fast-growing tumour, whereas most acoustic neuromas are slow-growing. The surgeons have decided not to operate until they know for sure that the tumour has grown rather than swelled after the non-invasive form of surgery I had at Leeds Gamma Knife Centre. This tumour is getting in the way of my ability to live as I want to and I almost feel like electing to have it removed. The right-hand side of my face has become numb and I can only eat properly on the left. My mouth does not close properly so I don't have as much control as before when I'm drinking or brushing my teeth. My right eye does not close properly, which interferes with my ability to read and write, two things I can still enjoy. My balance is affected and I cannot walk straight; I walked slowly anyway but now I look as though I'm drunk, tottering about.

I can't help thinking of how I would have been able to explore properly in the past. I would probably have been able to find evidence of the burial grounds or seen some of the grikes that my uncle mentions, which he thought might be responsible for the wood's survival. I saw some at Colt Park, formed by water rivulets channelling beneath the limestone escarpment and causing cracks to appear.

Tom Cooper claims that the presence of the burial grounds might have dictated the approach of the woodland authorities in these spots: the trees here appear to have been left alone, while those further up the hill have not.

Some of the ferns are going brown and yellow. They are beginning to die and soon the trees will lose their leaves. Ashes are among the first to shed their leaves, even though they are one of the last to get them. In one section a whole tree has fallen, lying prostrate in the grass. There are no leaves, so it has been dead for at least a year. It has probably been on the forest floor since last summer, exposed to wind and rain. One of its branches is pointing upwards, as if indicating the sky. It was at least a hundred years old, judging from the thickness of its trunk. I draw closer and see that the verdure coating it is lichens and mosses. The mosses are emerald green and they feel furry to the touch. They are soft and slightly damp when I stroke them, oily like feathers. When I look more closely I see that they resemble tiny fir trees and cover rocks and trees alike.

The name 'Rassal' is reminiscent of what is believed to be its ancient past, linking the woodland to its Viking history, as *Vikings in Scotland* attests: 'The presence of horses is also indicated by the names Rossal ("horse-field") or Hestaval ("horse-hill").' 'Rossal' is thought to be of Norse origin and the name is supposed to mean 'the care of horses'. Whether this is a reference to the fact that horses were kept nearby, I don't know, but it does seem that the wood was used by Vikings. Without horses the Vikings would never have spread as far as they did. Horses meant they could travel further and transport more goods. Odin kept the most famous horses, Sleipnir, daughter of Loki, in the form of a mare, and the stallion Svaoilfari, thought to have possessed amazing vigour. In Jesse Harasta's *Odin: The Origins, History and Evolution of the Norse God*, Odin is shown entering Valhalla on Sleipnir's back – she is unmistakable as she has eight legs. There is also a seventh-century stone image of Odin, this time on a four-legged horse leading warriors into battle.

There was a suggestion that the areas where there are no trees, just scrub and bracken, might have been cleared for grazing sheep or other farming. It might have been the Vikings who originally cleared the trees. Applecross, a sprawling seaside place nearby, is rich in Viking history. Their longboats would have had open access and would have been able to land easily. There is evidence that, after a tumultuous start, the Vikings began to integrate and become part of the local community. Helen Murchison upholds that Rassal was part of the vast Applecross Estate and it might have been belonged to Strome Castle in medieval times.

The Vikings who settled near here are reputed to have conquered Scotland between the ninth and twelfth centuries. They made good farmers and raised cattle, sheep and pigs. Barley and oats were grown along with other cereals. Burial grounds that have been unearthed indicate that many women came over, perhaps sent for when their men had done the conquering. These people may have been the ancestors of the families that settled near to Rassal. The area around it was occupied by crofters, who used the woodland for timber, shelter and to graze their animals. Their way of life was closely connected to their environment and continued so for centuries.

Tom Cooper states that rents were paid in food, since there is a record of cheese, butter and eggs being given to the landlord. Survival was uppermost and, as the crofters must have been very poor, eking a living from the land, Helen Murchison says it was in their interests to maintain the wood. In the 1700s, now owned by the MacKenzies, it flourished, but when the family died out it was sold to the Duke of Leeds. There were clearances from 1840 to 1860 and the indigenous population was forced out of the glens. Tom Cooper states that these evictions caused great hardship and strife and were reported in the local press. Deer parks were formed in their wake, for the wealthy to shoot at their leisure.

Regretfully, we decide to leave the woodland and walk back slowly the way we came, leaving the birds and trees to themselves. We have been here most of the day but it has passed so quickly that I hardly noticed. Helen Murchison tells me that on 16 September 2015 the *Sunday Herald* reported that Ash Dieback had been found in central Sutherland and in the forests west of Inverness, in the Great Glen and Wester Ross, in Morvern and down into Argyll. So I look for evidence of it in the trees but find nothing.

When I get back to our lodgings that night, I read John Burnside's poem 'September' which seems particularly apt:

> and listen, through the river of the trees,
>
> for something of myself that waits to come,
>
> as lyrical and poignant as the sound
>
> of little owls and foxes on the hill
>
> hunting for blood and warmth, in the yellow bracken.

This seems to capture the moment, as if it were written in Rassal Ashwood, and I go to sleep dreaming of foxes in the undergrowth.

*

It is not until May 2016 that I speak on the telephone to David Genney, the bryophtes, fungi and lichens policy and advice officer for Scottish Natural Heritage. He has been to Rassal Ashwood a few times and carried out some work there, but the area he has to cover is vast, the whole of Scotland. It is the first time I have spoken to Scottish National Heritage and David says I shouldn't compare the state of the ashwood before it was delisted and after. He confirms that the environment has always been a bit neglected and it is therefore no different from before. The ashwood hasn't lost any of its statutory protection and in a sense they've got the best of both worlds because they still have a say in the management of the wood, which has always been privately owned. It was apparently part of a wider move to make places more accessible to the general public.

David tells me that the landowner is encouraging grazing by cattle in an attempt to keep down the bracken. He adds that a few years ago a fence was erected at Rassal to promote regeneration and, as a result, the honeysuckle has overgrown. Too many grasses and plants have taken hold, which lichens don't like. On the outskirts of the wood the trees have more light and it is drier because there are no windbreaks. Scotland experiences quite mild wet winters and cool, wet summers, particularly in the west. Depending on the area of the woodland, conditions differ: in the centre of the wood it is humid so mosses and liverworts prevail.

I saw what I thought were sphagnum mosses growing on some boggy ground, looking like lots of minute ferns, so green they almost hurt my eyes. The peat they were on was darker and is quite abundant at Rassal. Sphagnum mosses were used to dress wounds during the First and Second World Wars because of their antiseptic properties. It is believed they were collected and sent to the south, where they had various other uses, as nappies, bedding and lamp wicks. They are still used by gardeners, who find them indispensable for the linings of hanging baskets and to pack plants for transport. Sphagnum mosses are very acidic and make possible the reconstruction of historic events because they preserve things very well. Years ago, at the British Museum, I saw 'Lindow Man', who was discovered by chance in a bog in Cheshire in 1984; he is thought to have been put there between 2 BC and AD 119. At the time, I didn't know that sphagnum mosses preserve bodies in bogs – apparently so effectively that it is difficult to date them.

The conditions at Rassal Ashwood are perfect for lichens, which favour shade and humidity, while mosses prefer the conditions just outside the wood. At Rassal there are two ravines, one at the south and one at the north, which bryophytes like. Rassal Ashwood is home to the largest collection of *Leptogium saturnium* in Britain, which David Genney photographed: it is growing on a rock and is black and curly, like ears, contrasting with the rock, which is white and porous. *Fuscopannaria ignobilis*, a type of lichen resembling small sea creatures, grows on the bark of trees and is common in Rassal too. It is known as a Shingle Lichen and is a pale orange.

At Rassal there are some *Lobaria pulmonaria*, or Tree Lungwort, which is used to treat respiratory symptoms. In *The Signature of All Things*, Alma comes across a book by Jacob Böhme, a sixteenth-century cobbler who believes that in nature we will find cures; each plant resembles the disease it is meant to cure or the organ it is intended to treat. He calls this 'the doctrine of all signatures' saying, for example, that tree lungwort resembles lungs. His logic was medieval and Alma did not take him seriously. Not everyone agrees, though, because a few years ago a French team were trying to harvest *Lobaria pulmonaria* in Scotland. There are many thousands of mosses and I can barely tell the difference between lichens and mosses. In *The Signature of All Things* Alma claims: 'Mosses hold their beauty in elegant reserve. By comparison to mosses, everything else in the botanical world can seem so blunt and obvious.' It is not until I look closely that I see the differences in the types of mosses and notice that they are all singularly beautiful.

David Genney is hoping that Ash Dieback won't arrive in Rassal, but I am pessimistic. Scottish National Heritage has not yet been informed that it is in Wester Ross. To begin with, the disease predominantly affects young trees – or, rather, older trees take longer to show signs of it and, since it is older trees that attract lichens, there will be no immediate change in them. Many species will be lost if *Chalara fraxinea* does arrive, although the remaining populations might move to hazel. There is some talk of sycamore replacing ash because it is quite similar: it has many of the same ecological characteristics as ash, such as the age and frequency of seed production, though it creates a much denser canopy, but it is currently treated in Scotland as an invasive non-native. The continuity of trees is important so once the ashes have gone, years from now, the woodland will never look the same again.

The person who knows the wood most intimately is, ironically, the last person I ring. It is July 2016 when I speak to Jenny Baker-Patch, who has lived on the farm nearest the wood for thirty-six years. At one time she was honorary warden at Rassal but restrictions on lone working meant she couldn't go to the woods by herself so she had to give it up.

At least twenty years ago there was some replanting of ash and rowan so I think that the lines of rowans were planted then. They are distinguishable from the ashes by their silvery pale bark. Jenny confirms that there has been no discernible change in the environment over the last two years; it looks more or less as it always did. Bracken grows faster than it can be trampled down and the cattle are allowed into the woods only at certain times of year.

Common Ragwort, a yellow-flowering plant that looks a little like tansy, grows here abundantly. The World Horse Welfare website reports that a horse can die suddenly as a result of ragwort poisoning. It doesn't stop flowering until around October and can grow to a height of up to two metres. The website tells me that it is a horrible way for a horse to die, and I wonder if ragwort grew here in the time of the Vikings. They would have known that ragwort is harmful to equines so it is surprising, perhaps, that Rassal Ashwood is named after horses.

Rassal would look bald without the ashes. The grikes would be exposed and the ravines denuded. The area has evolved in this way to support the trees or perhaps the trees have evolved because of the area. An excavation would surely take place, and who knows what it might reveal? Perhaps Viking remains or a burial from even earlier. There is no doubt that Rassal hides something beneath its thick cluster of ashes on the hillside.

Ash Dieback is encroaching on Scotland, and in 2015 there was an astonishing number of cases. The Forestry Commission maintains that it has jumped from five new cases in 2013 to 125 in 2015. The most northerly example of the disease is in Lairg, the most westerly on the Isle of Mull, and on the mainland it is already in Oban. Rassal Ashwood cannot avoid the disease because spores will be blown on the wind and Ash Dieback will develop in the trees. The Forestry Commission says that because more people have been looking out for the disease, more cases are reported, accounting for the increase. In 2015 statutory legislation meant that more experts became available to identify it.

The Forestry Commission believes that Ash Dieback has probably been in Scotland for a long time – it is impossible to say for how long. In *Ecology and Modern Scottish Literature* Louisa Gairn suggests that, in his poetry, John Burnside made 'an intuitive designation of some places as "sacred", a place of the dead – or of the devil – which must be left to nature.' Let us hope that others will share this sentiment and that the bones beneath will be left to rest in peace.

Shadows

M. R. James's *The Ash Tree* begins with a whimsical description of Castringham Hall. We get an impression of the house being grand, although it is portrayed as one of 'the rather dank little buildings'. In this story the tree is central to the plot: it shows the sinister side of the ash, portraying it as a portent of bad luck. This tree is fictional, of course, but there were many such trees up and down England. In this case, the ash is in Suffolk, an area that has been greatly affected by Ash Dieback in recent years.

The fate of the ash growing close to the old hall on its west side becomes entwined with that of generations of the Fell family, who own the hall. We are told: 'As you looked at it [the hall] from the park, you saw on the right a great old ash tree growing within half a dozen yards of the wall, and almost or quite touching the building with its branches.' Our curiosity is established because the tree is no longer there and we wonder why such a distinctive feature was removed. The story of the tree and the hall then unfolds and we understand more about this association. Sir Matthew Fell, then the owner of Castringham Hall, testifies that Mrs Mothersole, of the same parish, was a witch because he apparently saw her three times at night climbing the tree in her undergarment, cutting ash twigs and talking to herself. That was enough to condemn her to death despite her previous good character. In May, Sir Matthew and the vicar, Mr Crome, a close friend, saw some strange creatures moving in the tree and thought they had more than four legs. The next morning Sir Matthew was found dead in his bed. The window had been open all night. He seemed to have been writhing in pain before he died, and the two women who laid him out complained of swelling and aching in their hands and arms, which lasted for a few weeks. Sir Matthew was thought to have been poisoned.

Mr Crome used the Bible to make some predictions, which he recorded in his private papers: 'From Luke 13.7, "Cut it down"; in the second, Isaiah 13.20, "It shall never be inhabited"; and upon the third experiment, Job 39.30 "Her young ones also suck up blood."' In the lifetime of Sir Matthew's son, the death rate of animals was higher than average but, other than that, nothing remarkable occurred. His son, though, Sir Richard Fell, was the next to die in mysterious circumstances in the same room as his grandfather. The servants found him in much the same condition and everyone was baffled.

This story is about the malevolent creatures that lived in the bole of the tree, huge hairy spiders that crawled out at night and poisoned all the animals and people they could find. It is said by the Bishop of Kilmore, one of the visitors to the house, that Irish country people believe it brings ill fortune to sleep near an ash tree. In *The Spirit of Trees*, Fred Hageneder says, 'Ash trees, however, were not planted close to the house or crops, because their vigorous root system can damage stonewalls or hinder the growth of crops.' To have written this ghost story, M. R. James must have been aware of the negative connotations of the ash and its spiritual properties. He must have believed, at least for the purposes of the story, that they were thought by some to be trees of ill omen.

In *The Ash Tree*, the spiders that killed Sir Matthew Fell might have been the same ones that killed the animals and Sir Richard because spiders can live indefinitely – that is, they can survive for years. The spiders took shelter in the ash and made their nest in it, a bit like the tumour that has taken root in my brain. My surgeon says that an acoustic neuroma is caused by a mutation in a cell. If it had been left undisturbed, it would have begun to compress the other organs in my brain and would eventually have ended my life. The spiders inside the ash aren't supposed to be there because, over a period of time, they wreak havoc. I cannot help but think of my body as an ash tree that harbours an enemy within. In a way, I have to accept the darkness in order to recover. At times I feel sorry for myself, but that doesn't help anyone.

One night when I go to bed earlier than usual, because I'm so tired, I wake up in the small hours and cannot get back to sleep. Eventually I fall into a doze and begin to dream. I dream that I am giving birth, great pushes that hurt me and pants that cause waves of nausea. Instead of giving birth in the normal way, my stomach splits open and reveals giant spiders inside covered with dark fur, like the ones that have their nest in the fictional ash. The spiders crawl out slowly and, rather than three or four, there are many. They keep coming, a steady march of spiders that gradually surround me. I am rooted to the spot: I cannot run away because my stomach is disgorged.

I wake up bathed in sweat. Mike is sleeping beside me, oblivious of my distress. I am shaking like a leaf but I manage to turn on my bedside lamp and sit up. The lurid dream is still real to me but I realize there are no spiders; my skin is itching as if it is on fire. I am reassured when I feel that my stomach is flat and there isn't a large split, just an appendix scar from when I was ten years old. I take a drink and my breathing gradually returns to normal as I calm down. Mike wakes and tries to soothe me.

I am grappling with the damage that my illness has done and the dark themes of *The Ash Tree* seem to echo in my life. Apparently, when my brain was operated on it bled so much that it was difficult for the surgeons to see what they were doing. I have to find a way of coming to terms with the physical changes the tumour has caused and the psychological ones. I am afraid of dying, and it is easy to imagine danger lurking round every corner, waiting to seize and kill me, like the malevolent creatures that live in the ash of the story. The tumour still inside me, 'benign' though it is, or the ordeal I've undergone, could be the cause of my negative feelings. There is clearly something dark in me that needs to be expressed before it becomes bitter and distorted. I must be traumatized by what has happened to me, still coming to terms with my slowness and feeling frustrated every day. I rarely sleep a whole night through: I tend to wake in the early hours, then toss and turn until morning.

I read in the newspaper one Saturday morning a review of a book by Gavin Francis called *Adventures in Human Being*, about his experiences of practising medicine. He describes holding a human brain in his hands when he was a medical student. It is the first time I have ever read a description like this: 'It was heavier than I had anticipated; grey, firm and laboratory-cold. Its surface was slippery and smooth, like an algae-covered stone pulled from a riverbed. As I cradled the brain in my hands, I tried to think of the consciousness it had once supported, the emotions that had once crackled through its neurons and synapses.' This makes me feel strange: I don't like to think of the surgeons poking around in my brain. I prefer not to consider it even though my existence bears testament to the fact that they did. Gavin Francis views the brain objectively, which I am unable to do. My brain would resemble this one if it were removed; it would become like any other. My emotions would no longer be important and my consciousness would be as nothing.

All my memories are intact, for which I am very grateful. My perceptions are the same as they were before; I am still stubborn. An acoustic neuroma is also called a vestibular schwannoma and there is no doubt that my vestibular nerve has been affected. It controls balance and eye movements. In *The Handbook of Vestibular Rehabilitation*, edited by Linda Luxon and Rosalyn Davies, there is a reference to events that can trigger dizziness: 'i.e. head or body movements and, specifically, turning over in bed, sudden loud noises and also, certain visual environments, such as milling crowds, supermarkets, fluorescent lighting, or empty corridors'. I have felt dizzy because of all these events, but sometimes my head is not affected so I never know if these incidents will provoke an attack. If it happens, I just stand still until it has passed. I don't tell anyone: it's easier that way.

The darkness inside me is resentment that I am no longer able-bodied. The worst thing for me is not being able to walk wherever I like in the countryside. I ask myself, 'Why me?' I don't know why but I just have to get on with life. Things are not at all as I would like them to be but they are as they are. There is not much I can do about it, although there are some things I can do to improve matters: doing my exercises, getting used to functioning at a slower pace, not allowing myself to become so frustrated and learning to like myself. There are many people in the world whose plight is worse than my own. If I focus on myself, I become more upset. It is very common in someone who has been through a traumatic experience to be selfish so I try not to be. I am protective of myself because I know that I cannot do what other people can. A small thing, like getting on and off a train, is difficult for me now but for most people it is straightforward. On local trains, I can manage on my own by holding on to a handrail and stepping down carefully. On intercity trains, though, I need help getting on and off so I have to rely more on able-bodied people than I did before. It is a matter of trust, being part of society and learning to let family and friends help me.

Shelter

On a fresh September morning in 2015, when the first leaves of autumn crunch underfoot, I sit in the shade of a fat beech in St Pancras Churchyard. It is quiet and calm, a place to gather your thoughts and listen to the birds competing with the roar of trains and planes, and traffic screeching. I listen intently to the distinctive two trills of what I believe to be a bullfinch; it pauses, then calls again, the same each time, like a broken record. I don't see it but bullfinches like ash. After a while I hear only the birdsong, though a train snakes by as it leaves St Pancras station, its long roof visible above the wall. Bullfinches eat ash seeds and they are sometimes to be found on an ash branch eating the keys. It is rare to see a bullfinch because they are quite shy birds and there aren't many of them.

I am here to see Hardy's Ash, which is one of the tallest and finest trees I have seen. It bends its many boughs to the earth in a courtly dance, swaying slowly with the occasional breeze. It is encircled by a low privet hedge and black railings to keep eager sightseers off the headstones. More and more, I find myself in graveyards because they provide a soothing space. Being close to the dead makes me feel more alive.

The thick roots of the ash have grown over the headstones that are placed on their sides, like dominoes lined up. They are built on a slight hillock and consequently they resemble the top of a pineapple. This could relate to the 'Many hundreds of coffins, and bones in huge quantities' that, according to Claire Tomalin, in *Thomas Hardy, the Time-Torn Man*, were apparently piled up here. Iphgenia Baal's novel *The Hardy Tree* claims that 432 headstones were put around the tree in twelve layers. At that time, the tree was so small that the stones made it seem even more diminutive. The roots of the ash have wrapped themselves around the graves and in places they appear to be red or pink, like blood.

Thomas Hardy intended the ash to guard the hundreds of headstones that had to be moved from their original resting place in the old St Pancras Churchyard to make way for the Midland railway line that would terminate at the station. At the time, Hardy, still a relatively young unmarried man of twenty-six, was working as an assistant to a successful London architect, Arthur Blomfield, and it is surely a sign of the esteem in which Blomfield held him that Hardy was given the task of overseeing the moving of the graves. The Bishop of London had asked Blomfield to perform this duty after the public became aware of what was happening to the cemetery and believed that the bodies interred there were in danger of being defiled. Actually, the bishop appears to have been related to Blomfield, so it was a clear case of connections. However, the bishop knew that Blomfield would do the job properly and Thomas Hardy was instrumental in putting the headstones here, an unlikely task for a man who had recently decided that he could not accept the established doctrine of Christianity. The tree was perhaps self-seeded because, as maintained in Baal's novel, 'no one is able to explain its arrival, but it is planted in soft ground', so it was ideal for digging a pit around. Even then, Hardy must have believed that trees had spiritual properties and that underneath an ash tree was a good place for the headstones. Later, like Mary Webb, Hardy became a pantheist, who believed it was essential to be in touch with the natural world for the gods to reveal themselves.

At the time, St Pancras Churchyard was a dismal place, encircled by a workhouse and poor housing, not at all somewhere you would want to be at night, which was when Hardy had to be there. The churchyard, once surrounded by open fields, became the scene of coffins being dug up behind boards. The work was carried out under cover of darkness, to minimize disruption and keep out prying eyes. It is no wonder that Hardy became depressed during the autumn and winter of 1866. He and Blomfield witnessed one of the coffins fall open to reveal one skeleton and two skulls, which must have been terrifying, especially if the bones were lit by flare lamps. It was to be Hardy's last winter in London. He wrote 'Neutral Tones' during his London period, a poem about the break-up of a relationship in which ash leaves represent the lifeless nature of the relationship: 'They had fallen from an ash, and were gray.'

The Hardy Tree must have seen many changes. It must have been there when so many people flooded from the country to the city to work in new factories, during the Industrial Revolution. It must have seen Queen Victoria's death in 1901. It was there through both world wars and must have been the scene of many tearful moments for women and men. The buildings and railway lines have evolved around it, noisily encroaching on its space, but it has stood firm throughout all this, quietly sheltering the headstones that were moved under it all those years ago. In her novel, Baal muses that 'It is in 1991 that the grey ash finally scales the cemetery walls for the first time to see the host of trains, cranes and automobiles.' It is tall and strong, though hemmed in. I feel a little sorry for it because it has never been allowed to grow without being useful, but I suppose that is the lot of many trees.

The lichen-covered branches have expanded to support the weight of the tree's many graceful limbs, which have been allowed to grow as it likes. Some groups of tourists are visiting the tree, wandering around and trying to read the inscriptions on the stones. A few scars where branches have been lopped have left knobbles where, in a face, eyes would be and, beneath, a gnarled lump in the form of a nose. A woman behind me says it resembles 'one of the Ents from Lord of the Rings'.

I shared a room in the rehabilitation centre with a young girl who had lost her leg in a road accident. She was depressed and coming to terms with what life would be like without her leg, but she never lost hope. A human has two legs, but a tree has many limbs and can grow more. However, if we lose a limb, we have the intelligence to develop prosthetic arms and legs.

On the side facing away from the church, the view is of a long arm halfway up the trunk that has grown like the leg of a ballerina in a *plié* pose. The body of the tree leans forward in a gesture of grace and compassion, honouring the dead. I can just make out the name of one Thomas Morton who died in June of an unknown year, his bones disinterred 150 years ago but his life commemorated here under the Hardy Tree. I wonder whether he would be glad to know that his memory is held sacred and guarded under this magnificent ash. Some of the stones have cracked under the weight of the tree, tucked so close under it that half of them have disappeared. I slip round the narrow gap between bush and balustrade to peer closely at the name on a chipped, mottled stone: 'Susan Rogers, 8th February 1883' – I lean closer over the etched letters – 'aged 87'. Her stone was put there after the original cemetery was moved to make way for the station. They didn't leave her for long in her first grave before removing her in the name of leisure: the great steam railway ate into many community and sacred places in bringing the seaside and the countryside within the sights of unhealthy city dwellers.

Nowadays, the trees within London are threatened. According to the report *Tree Diseases In London*, published by the City of London in 2012, ashes form 3 per cent of tree cover, which doesn't sound much but their loss will drastically alter the townscape. With them will go the benefits they offer. It is well known that trees supply oxygen and remove carbon dioxide, thus making environments where trees are attractive to people. As maintained by *Tree Diseases in London*: 'The roots of trees extract moisture from the ground, with surplus leaving as water vapour in the sky.' When it rains, trees help drain the water by absorbing some in their roots, making pavements easier to walk on. The leaves of a tree provide shade from sun and rain and we often shelter under them. Birds sing in them, which relaxes us, so there are psychological benefits as well. As *Tree Diseases in London* says: 'Bird song comes about only because birds have places to rest, feed and breed.' Trees are a form of shelter for birds, protecting them from the elements and bestowing on them a home. The impact of Ash Dieback in London will be minuscule, but it will be one less species to provide variety and it will be another threat to trees.

Far up north, in Yorkshire, on a hot June afternoon in 2016, we have to drive with the windows wound down to allow the breeze to fan us. I enjoy the rare feeling of heat. We park down a lane in the village of North Grimston, in front of a low wall separating us from a field. I notice a sign warning of a bull and protective cows. Our big dog is a rather enthusiastic labradoodle and cows are often disturbed by him. I am determined to see the ash, so there is nothing for it but to skirt round the edge of the field. The Settrington Ash is the prize I am not prepared to forfeit. In the event, the cattle are round the corner in another part of the field, too busy munching grass to take any notice of us.

I am nervous and try to hurry as the grass brushes against my legs. The dog is too busy panting in the heat to bark. We climb across a stile and into the next field safely. I breathe a sigh of relief. I spot the ash from a distance and I can hardly see the trunk for the dense foliage: the branches are laden with dark green leaves that are so heavy they nearly reach to the ground. Lots of seed pods dangle from the tree, paler than the polished leaves, a lighter green than the tree itself, almost yellow. The great hole in its side is a surprise because the boughs are full and lush. It was probably hit by lightning, which caused a wound or crack to open and the hollowing to begin. A hollow ash is often a sign that the tree is ancient and it may have been like that for at least a century (judging from its width). There is less weight to support so leaves and seeds can grow in abundance.

The hollow of the tree affords a roughly circular space four or five feet in diameter. There is just enough room for the two of us, Mike and me, to stand up and eat our sandwiches, albeit in rather close proximity. It does not live up to the fantastical proportions of the 'ash-dotterel' described by John Clare in his poem *The Hollow Tree*:

> How oft a summer shower hath started me
> To seek for shelter in a hollow tree
> Old huge ash-dotterel wasted to a shell
> Whose vigorous head still grew and flourished well
> Where ten might sit upon the battered floor
> And still look round discovering room for more
> And he who chose a hermit's life to share
> Might have a door and make a cabin there –
> They seemed so like a house that our desires
> Would call them so and make our gipsey fires
> And eat field dinners of the juicy peas
> Till we were wet and drabbled to the knees
> But in our old tree-house rain as it might
> Not one drop fell although it rained all night.

Like John Clare's tree, the Settrington Ash offers protection from the elements. It is cosy inside and I hide for a while, standing still. I want to blend into the environment, to focus on the tree and lose myself. In the winter I had another MRI scan and was told that my tumour had reduced in size. It meant an end to all the worry and uncertainty. It meant that I could get on with my life. Despite the good news, I am still adjusting and getting over the upheaval. I don't think I can ever return to how I was before because I'm not the same. I still have to have scans every six months but that is routine now. The psychological scars are still there and I don't think they will ever go away. Hopefully, I will learn to see the illness as something that happened to me in the past. I know that I will never again take anything for granted.

When you knock on the side of the trunk it is a gentle sound, but reverberates like the hollow inside of a guitar. Some guitars are made of ash, and part of the trunk forms a bowl shape. I've never seen a trunk from the inside. I've never been inside a tree before. The smooth inverse shape is hard to get used to but I think of it as a mould for making clay figures seen from the inside, so that I *feel* or imagine the indents and furrows of the trunk rather than observing them from the outside, almost as if they are the imperfections of my own skin. I feel strangely at peace with myself and the world that passes me by, though not that much passes here, just birds, bees, insects and sheep. It is as if the walls of the trunk, thin though they may be, are protecting me from harm. I'm sure it's all in the mind, but Robert Macfarlane suggests in *Mountains of the Mind* that our behaviour changes in response to where we are. Our responses to (and therefore our perceptions of) land-scapes are culturally devised, and he suggests that our behaviour changes wherever we are.

I sit under trees at every opportunity. I dream when I'm under the boughs of a tree and my thoughts can drift and deepen. When I was a toddler, Uncle Gerald gave my sister and me a Wendy house for Christmas, and I remember sitting inside it listening to the adults talk elsewhere in the room. The wooden house was a refuge and a place of safety from prying eyes. Now I find myself whispering to Mike so as not to disturb the smaller animals that might be in or around the tree. I try to move silently and make as little disruption as I can, as if they can pretend that I'm not really here. That I am inside the trunk is an immutable fact, an experience I can never forget, partly because I don't want to.

There are lots of tiny insect holes and I observe two flies disappearing into them. I've never seen that before and I am fascinated: the flies must live in the holes. Given the short length of time that flies live (six months at most), it is likely that the holes have been used by generations of flies. They might be *Psyllopsis discrepans* or *Psyllopsis fraxini*, which are very similar. The British Bug website, which I look at that evening, says *Psyllopsis fraxini* are easily distinguished by the 'dark clouds around the apex of the forewing' but they all looked the same to me. My fleeting impression, though I could be mistaken, is of a pale green back so it looks like *Psyllopsis discrepans*, also known as Cottony Ash Psyllid. It is unlikely that these flies will exist at all once Ash Dieback has arrived because they prefer ash to all other species.

After a short time, I take on the form and perspective of the tree, looking at the field from its point of view and expanding into the space, just as we did in primary-school physical-education lessons when our teacher asked us to imagine we were trees and to sway as a tree would. I found those lessons thrilling, because at school we were so rarely asked to imagine anything, and for a child with an overactive imagination, like me, it was a licence to go wild: I would throw myself around the school hall in imitation of a tree in a thunderstorm. Now I can hear a branch thud against the base of the trunk in the wind. The ash used to be called the 'widow-maker' because its limbs are prone to fall unexpectedly on passers-by. Of course, that was in the days when people walked everywhere. Now most people drive and the only precaution we need against falling boughs is insurance.

Lime green and pale yellow, the leafy *Xanthoria parietina* lichen spreads on the grey of the bark, like decoration. I recognize this lichen because it is very common in

England and grows in abundance on this ash, crusty and brittle, like a witch's fingers. It apparently likes nutrient-enriched bark. Some branches are completely covered with it and curl upwards like the tentacles of a large sea creature. The internet reveals that it was used as a cure for jaundice because of its yellow colour. Some branches have fallen into the grass, and I notice lots of small dark pebbles growing on them. They are known as King Alfred's Cakes because they are said to resemble the burned offerings that came out of the oven when the king fell asleep. Their real name is *Daldinia concentrica*. I look more closely and see a beetle crawling on the bark with a fawn head and bottom, lighter than its black back. I think this is a Cramp Ball Fungus Weevil, which lives on the dead wood of ash. Its larvae develop in the dark inside King Alfred's Cakes, so they will survive as long as there is dead ash wood. Once the ash has gone, they will die because they live only on ash.

After the decline of the elm, the ash became home to more varieties of lichen and other dependants. Beetles, mosses, lichens and fungi thrive on ancient ashes, with their hollows and decaying branches. The older the tree, the more dependants it attracts. Some moths feed exclusively on ash, such as the Tawny Pinion, and others that feed mainly on ash, including the Barred Tooth-striped moth and the Ash Pug. I communicate with Dr Charles Fletcher, a moth expert who lives near me, and he says that Ash Dieback is a constant worry: 'A lot of invertebrate species depend on it,' he pronounces. I see a photo of the Tawny Pinion on a website and it resembles a piece of wood, easily camouflaged by the tree it is on, which must be an ash. Charles says it is present at a low density across the county. Very few moths feed exclusively on ash: some feast on honeysuckle, privet and lilac. The Ash Pug eats only ash, as does the Centre-barred Sallow, which Charles says is a common moth. If you look closely at moths they resemble dinosaurs, small reminders of our past. The wings of the Ash Pug are grey or dark brown, so it is not beautiful but it is compact, almost perfect. Other moths that depend on ash are the Coronet, Privet Hawk, Lunar Thorn and the Ash Bud. Charles suggests that several other species may depend on ash for their survival. In his report in the *Journal of Ecology*, Dr Peter Thomas records more than a hundred insect species that depend on ash. Some will move onto other trees but a small number will become extinct.

Outside sounds are muffled inside the tree and the walls of the trunk are solid, if thin, worm-holed vertical ridges wavering from root to roof. There is a feeling of safety, of being protected from outside forces, like in church. Inside the trunk, sound is muted and I am surrounded by wooden things, but there are no other people here now – Mike has left me alone – which is different from a church, and no one is telling me what to think, like a vicar. This is in keeping with John Clare's belief that the hollow ash in the poem would shelter and protect him, and it proves Jonathan Bate's statement in *Song for the Earth* that Clare, in writing 'The Hollow Tree', was looking for a home in the natural world. For an isolated moment Clare lives inside the tree: like a tortoise in its shell, he occupies the enclosed space for a fragment of time and feels that it is his refuge. He needs a calm place as the hustle and bustle of the outside world are too much for him. For me, the tree affords a safe place to be, somewhere I can contemplate things and my connection to them.

Paul Farley, in his introduction to *John Clare*, containing selected poems, calls him 'the great poet of dwelling and displacement' and says that 'Clare's early life coincided with the greatest spatial upheaval that the English countryside – and townscape – had seen in a millennium.' I don't think this tree is old enough to have seen the Enclosure Act in 1709, which got rid of the open land and meant that fields belonged to certain people, which they always had but others could roam freely. There is no doubt that there was an ash in this field, an antecedent of this tree, which witnessed it.

There is a gaping hole at head height for me to peer through. It is airy inside the tree, with a small fissure under the base, as well as the door, so when the wind rushes through, as it is does now, the hollow tree becomes a musical instrument that whines and reverberates to the rhythm of its airy bow.

This trunk cannot be repaired. Nor do I want it to be. In here I am hidden from the cows in the next field and sheltered from the wind. I share my new home with a furniture beetle and a number of spiders, dead and alive, that hang off the many gauzy webs. Sheep clearly shelter here, because there are skeins of white wool stuck to the inside, both old and new, forming a soft down or insulation. I can imagine the sheep taking shelter in a storm, all vying for space, which is at a premium.

A number of bulbous feet have spread their calloused sinews wide to support the tree, their toes curling deep under the earth of the wolds. The tree is squat and portly, at least four metres in girth, and it has held its ground well. I stand in the doorway as rain begins to patter on my wooden shell, surveying the field through the tree's low-hanging boughs covered with mustard-coloured lichen. They are bowed under the weight of their yellow keys and vivid green leaves. The hollow offers only partial shelter because the interior tapers up to a funnel, like a chimney, where a giant red fungus is perched, perhaps marking the tree as a dwelling for the little people. The wind is rising and begins to swirl round the tree, creaking its heavy old body. It is about five metres away from a mature mixed boundary hedge that shelters it from the worst of the gales.

There are more ash trees in the overgrown field, clearly a sheep field left to grow its rich grasses ready for its occupants to return in the autumn. The other ashes, hanging over the silty stream of Settrington Beck, are taller and darker in appearance. I wonder whether their proximity to the water has helped their roots to swell and grow faster, or whether they are a different strain of ash. They must be waterlogged for most of the year, while the hollow tree is in a dry patch, on slightly higher ground. The soil beneath the branches is cracked after the warm summer, but the base of the tall dark ash is rooted in the stream where watercress grows.

I can hardly bear to leave my new home, but the dog is getting impatient, tugging at his lead and pulling Mike into the beck. We are swept across the coarse grass, the thistles and the dock leaves. This moment heralds a glowing summer and my spirits are high. I am comforted by the discovery of this new place of shelter, a shell that I may think of as a home for my imagination to rest, a woody space to retreat to in the landscape of my mind.

I'm not the only one to regard an ash as a form of shelter. According to *A History of the County of Hertford*, written in 1908, the abbot of St Albans held his courts under the great ash in the grounds of the abbey. 'For the trial of all crimes, pleas, and plaints the townspeople had to go to the Hundred Court of the Abbot's Liberty, which was held before the steward of the liberty under the great ash tree in the courtyard of the abbey.' This took place sometime in the sixteenth century although I am not able to find an exact date. Apparently, courts were held under ashes in Sussex and Hampshire as well. This was probably a hangover from the Anglo-Saxon practice of holding them under an ash, since their deity, Woden (like the Norse Odin), gained greater wisdom from hanging himself on the ash, then coming back to life in a miraculous form of reincarnation, similar to Christ's resurrection. This would suggest that the canopy of an ancient ash offered not just shelter but that an air of solemnity and wisdom was associated with it.

For urban and country dwellers alike the ash, and trees in general, provide a sense of place. They help us to orient ourselves. The ash offered me shelter at the time in my life when I needed it most. I used the idea of it to hide from harm and I know that I can always return to the hollow tree in the near future. The danger is that ashes will become so rare that they will be only a memory, and young children won't remember them at all.

Resistance

March of 2016 is an unusually mild month and the birds are confused because it is often cold at this time of year. The Nornex Consortium has had a breakthrough: an experiment has enabled them to identify potential ashes that have tolerance to the disease. In Snorri Sturluson's *Voluspa* the Norns are three female giantesses, Uror, Skuld and Veroandi, who guard the Well of Fate, also called the Well of Uror, which is found at the foot of Yggdrasil. According to legend, they took some water from the well and poured it over the roots of the tree to stop it rotting. The Norns were weavers and it is believed they wove fate, determining what would happen.

Allan Downie is an affable, bearded Scotsman, who came out of retirement to lead the fight against Ash Dieback. He is in an office with a computer in front of him because we are speaking on Skype. There are rows of neat boxes behind him, which I feel as though I can almost touch and it is frustrating not to be able to do so. I am obliged to stare at Professor Downie all the time and cannot move much because he may not be able to see me. I don't like looking at the image of myself on the screen so I make it as small as possible and try to ignore my constant blinking and the way I scrunch up my forehead. I have developed a facial weakness that has caused me no end of problems over the last few months, particularly my right eye.

Dr Kjaer of the University of Copenhagen used Danish ashes for this experiment because it was thought more likely that they had been exposed to the disease: Denmark has had it for much longer than Britain. Professor Downie had the idea of using a technique called associative transcriptomics to find out the differences between trees that are tolerant to the disease and those that are suscept-ible to it. Ribonucleic acid is single-stranded, unlike DNA. It is an omnipresent group of large biological molecules that perform many essential functions in the coding, decoding, regulation, and expression of genes. RNA and DNA constitute the two major macromolecules, signifi-cant for all life forms that we know about.

In this method genetic markers are correlated with those trees that are tolerant to the disease, so all the mark-ers across the genome were looked at; the same was done with trees that looked as though they were susceptible to the disease. Initially, this all took place in Denmark because it was vital that the trees that were being experimented on truly had low susceptibility to the disease. It could not be done in Britain where too few trees have the infection as yet. It was not known whether the trees would be tolerant to the disease so they would give a good indication to help make predictions and to find out which genetic markers correlated with tolerance. The Nornex Consortium will be able to assess how good the predictions were. However, another cohort of trees has to be assessed to make sure that the results of the experiment are sound. Ian Bancroft and Andrea Harper in York have acquired trees from all over the UK to use in the new experiment, some of which are susceptible to the disease and some of which are resistant.

Professor Downie makes the distinction between trees that are tolerant and trees that are resistant: 'In a resistant tree you would expect that the fungus doesn't grow at all but on tolerant trees the fungus grows and forms some spores on the leaves of those trees,' he says. If the fungus learns to live on the leaves without killing the tree, the tree is tolerant, not resistant.

In Ashwellthorpe Lower Wood, most of the ashes are suffering from Ash Dieback now, apart from a few mature ones, one of which Anne Edwards calls Betty. She has collected lots of seeds from Betty and hopes to use them. There is a possibility that the seeds Betty produces have a higher likelihood of being tolerant. Professor Downie says that scientists are beginning to understand the pathogen better but there is still a lot of work to do. Saplings resistant to *Hymenoscyphus fraxineus* would be very expensive to screen using associative transcriptomics, and he is hoping it will be possible to change the DNA-based markers into RNA markers, but it would need more research. It doesn't make sense to screen for tolerant trees if the incidence of tolerance is high, but if just a few trees are tolerant, it will be important to identify them. Results from the experiments show that more trees than expected have one of the markers that predicts tolerance to the disease, but Andrea and Ian don't yet know whether this means that trees in certain areas will have more tolerance to it. Professor Downie estimates that over the next ten or twenty years we'll probably lose at least 80 per cent of our ashes to Ash Dieback.

A particular ash in a city centre could be saved if it is injected with fungicides or if Bio-char is used to alter the composition of the soil, reportedly successful in some cases. Bio-char is basically a form of charcoal rich in carbon. It wouldn't be possible to put it into all the soil for a wood or in a hedgerow as neither fungicides nor Bio-char are of much use in the countryside. There is a possibility that saplings planted now with more tolerance to the disease could eventually take over from the ashes that will die: ash grows very fast so that is a serious probability. It is critical from a scientific perspective to monitor the ashes planted with more tolerance to the disease but Professor Downie thinks that if we were to plant the seeds from trees like Betty, which seem to be more tolerant to the disease, we would increase the percentage of the ashes surviving.

Mike has seen a white crow in our local park and over the fields where ashes grow in a hedgerow. It looks, sounds and behaves like a crow, and has mated with a black crow, but it appears to be an albino, or perhaps leucistic, surely a rare and unexpected sight. When he took a photo of it flitting through the trees above him, it did not show up on the image, but its companions did. Like a ghostly shadow, it faded into the paleness of sky and cloud. I have not seen the bird, though I have been in search of it many times. No one we know has seen it either, so Mike is the only one who can bear witness to its existence. I poke fun, saying his eyesight isn't so good. I could doubt the veracity of his observation, but I know what he says is true: his delight is infectious.

Albinism and leucism, caused by lack of melanin pigment in the eyes, skin and hair, can occur in birds and fish as well as mammals. But it is rare, and most white crows end up in zoos because they are more visible to predators and therefore more vulnerable. It strikes me as a fearful metaphor for what may happen to the ash: it may become so rare that it can be cultivated and protected only in controlled conditions. We may have to take our grandchildren to see ashes in special reserves. If the ash is to survive it will have to be carefully monitored, just like in a zoo. It will survive, hopefully, but it will be rare, nothing like as abundant as it is now.

As Jonathan Bate recalls in *Song for the Earth*, Rousseau said that 'Conscience' is the 'voice of nature'. We value nature for the very reason that we are destroying it: the more we 'tame' nature in our everyday lives, the more we value 'wild' nature in our leisure time. My intention is not to mourn the past or lay blame at the feet of all consumers, industrialists and politicians, simply to reconstruct a part of the ash's story and evolution. However, it is incumbent on us all to ask what we can realistically do to support it. This book is a rallying cry to pick up your walking sticks, pens, paintbrushes and cameras, then record and enjoy what we have while we still have it.

Notes

The Blight of Ashwellthorpe

7 Wood. Blyton, E. (2015), *The Enchanted Wood*, London: Hachette, p. 5.

24 Soul seers. Schmitt, J.-C. (1998), *Ghosts in the Middle Ages*, Chicago: University of Chicago Press, p. 142.

The Science behind Ash Dieback

29 Magnified. http://www.psmicrographs.co.uk/ash-tree-pollen-grains-fraxinus-excelsior-/science-image. Accessed October 2013.

32 Student. Article: 20 August 2013. www.examiner.co.uk/news/west-yorkshire.

32 Arrested. www.ohio.com/news/break-news/group-unable-to-spare-ash-tree-in-highland-square. Accessed October 2013.

Secrets

47 Excavations. Batty, A., and Crack, N. (2009), 'Is Gauber the site of the Hermitage? A study by Arthur Batty and Noel Crack, compiled by Anita Batty', http://www.ingleborougharchaeology.org.uk/Archive. html

51 Uttley. Uttley, A. (1931), *The Country Child*, London: Puffin Books, p. 82.

Bread

56 Ivy. Browning, R. www.poetryfoundation.org/ poems-and-poets

59 Baking. Farquharson, B., and Doern, J. (eds) (2000), *Shops, Trades and Getting By*, Branscombe Project, pamphlet.

64 Fires. https://allpoetry.com/The-Firewood-Poem. Harmer, Ralph (1995), *Management of Coppice Stools*, Forestry Commission, p. 2.

Ancestors

73 Court. Website: www.discovery.nationalarchives. gov.uk. manor of windermere. Accessed October 2013.

74 Clare. http://www.poemhunter.com/poem/may-2/. Accessed October 2013.

75 Livestock. Bonfils, Marc, Fodder Trees in Temperate
Climate, www.permaculturefrance.org. Accessed
October 2013.

75 Trees. Rotherham, I. (ed) (2013), *Trees, Forested
Landscapes and Grazing Animals*, London: Routledge, p.
127.

76 Novel. Walpole, H. (1931), *Judith Paris*, London:
Macmillan, p. 140.

81 Painting. www.tate.org.uk/art/artworks/carrington-
farm-at-watendlath. Accessed October 2013.

84 Heisenberg. Heisenberg, W. (1958), *Physics and
Philosophy: the Revolution in Modern Science*, New York:
Harper & Bros., quoted in Robbins, J. (2012), *The Man
Who Planted Trees*, New York: Spiegel & Grau, p. 160.

88 Myths. https://en.wikipedia.org/wiki/Ásatrúarfélagið.
Accessed September 2013.

89 Protectors. Hageneder, F. (2000), *The Spirit of Trees*,
Edinburgh: Floris Books, p. 113.

90 Sagas. Hollander, L. M. (1962), *The Poetic Edda*, Austin:
University of Texas Press, p. 4.

Ash Fires

98 New Forest. Hooke, D. (2010), *Trees in Anglo-Saxon
England*, Woodbridge: Boydell Press, p. 132.

99 West Saxon poem. Hooke, p. 60.

107 Aesclings. Hooke, p. 201.

108 Types. 'The potential ecological impact of ash dieback', Joint Nature Conservation Committee, January 2014, jncc.defra.gov.uk

108 America. Little, C. E. (1995), *The Dying of the Trees*, New York: Penguin Books, p. 124.

110 Flying machine. https://airandspace.si.edu/exhibitions/wright-brothers/online/fly/

111 Charcoal. Hartley, D. (1939), *Made in England*, London: Eyre Methuen, pp. 46, 47.

113 River. Hooke, p. 201.

Life Cycle

118 Die. Wilks, J. H. (1972), *Trees of the British Isles in History and Legend*, London: Frederick Muller, p. 140.

118 Rupture. Frazer, J. (1994), *The Golden Bough*, London: Chancellor Press, p. 682.

119 Warts. Wilkinson, G. (1978), *Trees in the Wild*, London: J. Bartholomew, p. 27.

119 Bark. Hageneder, p. 112.

119 Solomon. Hooke, p. 77.

120 Anglo-Saxon poem, Gnomic verses, lines 25–29: Mackie, Exeter Book, II, pp. 32–35. Quoted in Hooke, p. 77.

122 Life. Hollander, L. M. (1962), *The Poetic Edda*, Austin: University of Texas Press, p. 50.

Spring

127 Amber. Webb, M. (1920), *The House in Dormer Forest*, Fairford: Echo Library, p. 103.

128 Biography. Barale, M. A. (1986), *Daughters and Lovers*, Middletown: Wesleyan University Press, p. 20.

129 Downfall. Webb, M. (1924), *Precious Bane*, London: Jonathan Cape.

132 Tale. Byford-Jones, W. (1948), *Shropshire Haunts of Mary Webb*, Shrewsbury: Wilding, p. 24.

133 Dependence. Coles, G. M. (1978), *The Flower of Light*. London: Duckworth, p.54.

133 Pantheist. Coles, p. 66.

135 Owners. Webb, M. (1917), *The Spring of Joy: Poems*, Gloucester: Dodo Press, p. 21.

138 Embankment. http://www.caughtbytheriver.net. Accessed 2012.

140 Wilderhope. Byford-Jones, p. 28.

141 Housman. www.poetryfoundation.org/ poems-and-poets/

142 Witchcraft. Hughes, J. (1977), *Shropshire Folklore, Ghosts and Witchcraft*, Shrewsbury: Westmid Supplies, p. 51.

Isolation

150 Seclusion. Wilkinson, G. (1978), *Trees in the Wild*,
London: Bartholomew, p. 25.

151–2 Moss. Gilbert, E. (2014), *The Signature of All Things*,
London: Bloomsbury, p. 199.

154 Dendrochronological. Peterken, G. (2008), *Natural
Woodland*, Cambridge: Cambridge University Press,
p. 376.

157 Name. Graham-Campbell, J., and Batey, C. E. (1998),
Vikings in Scotland, Edinburgh: Edinburgh University
Press, p. 207.

160 Lodgings. Burnside, J. (1995), *Swimming in the Flood*,
London: Cape Poetry, p. 47.

163 Lungs. Gilbert, E. (2014), *The Signature of All Things*,
London: Bloomsbury, p. 235.

165 Ragwort. www.worldhorsewelfare.org. Accessed
January 2016.

166 Sacred. Gairn, L. (2008), *Ecology and Modern Scottish
Literature*, Edinburgh: Edinburgh University Press,
p. 184.

Shadows

169 Buildings. James, M. R. (2007), *Complete Ghost
Stories*, London: Collectors' Library, pp. 52, 53, 59, 65,
68.

171 Crops. Hageneder, F. (2000), *The Spirit of Trees*,
Edinburgh: Floris Books, p. 113.

174 Brain. 'Adventures in medicine: I journey through the
body every day'. www.theguardian.com/books.
Accessed May 2015.

174 Corridors. Luxon, L., and Davies, R. (eds) (1997), *The
Handbook of Vestibular Rehabilitation*, New Jersey: John
Wiley, p. 32.

Shelter

180 Coffins. Tomalin, C. (2006), *Thomas Hardy, the Time-
Torn Man*, London: Viking, p. 109.

181 Self-seeded. Baal, I. (2011), *The Hardy Tree*, London:
Trolley Books, p. 20.

182 Lifeless. https://www.poetryfoundation.org/poems-
and-poets/poems/detail/50364. Accessed October 2015.

182–3 Walls. Baal, I., p. 54

184 People. Keen, J. (2013), *Tree Diseases in London*, City of
London Corporation, pp. 6, 8.

187 Hollow. Farley, P. (ed) (2007), *John Clare*, London:
Faber and Faber, p. 51.

189 Behaviour. Macfarlane, R. (2008), *Mountains of the
Mind*, London: Granta Books, p. 18.

189 Insect. www.britishbugs.org.uk/*Psyllopsis fraxini*.
Accessed June 2016.

191 Moths. http://ceredigionmoths.blogspot.
co.uk/2016/04/tawny-pinion.

193 Statement. Bate, J. (2000), *The Song of the Earth*,
London: Picador, p. 153.

193 Upheaval. Farley, p. xxi.

196 Shelter. http://www.british-history.ac.uk/*A History of
the County of Hertford* (volume 2 (1908), pp. 477–83.).
Accessed December 2015.

Resistance

206 Conscience. Bate, p. 35.

Acknowledgements

A lot of people have contributed to this book, not least my family. I am indebted to my sons, Daniel and Joel, and to my parents, Janet and Peter Samson. My editor at 4th Estate, Louise Haines, has read every chapter and her contribution was invaluable. My agent, Kirsty McLachlan, has read every draft and been very encouraging. I am grateful to the following people for giving their knowledge and expertise so willingly: Jenny Baker-Patch, Lizzie Bear, Peter Blyth of the National Trust, Ian Brittain of the Food and Environment Research Agency, Peter Carty of the National Trust, Steve Collin of Norfolk Wildlife Trust, Tom Cooper, Ivor Dowell, Maree Dowell, Allan Downie, Anne Edwards, Charles Fletcher, David Genney of Scottish National Heritage, Ted Green, John Greenslade, Ray Hawes of the National Trust, Alistair Heath of the National Trust, Leeds Beckett University,

Daniel McLean, Graham Mort, Brian Muelaner of the National Trust, Helen Murchison, Colin Newlands of Ingleborough National Nature Reserve, Maurice Pankhurst of the National Trust, the late Richard Richardson, Shaun Richardson, and my surgeon, Mr Kenan Deniz.